# Accident Prevention and Investigation

## A Systematic Guide for Professionals, Educators, Researchers, and Students

*Koji Fukuoka*
*BCP & BCM Consulting,*
*Tokyo, Japan*

WILEY

Published by John Wiley & Sons, Inc., Hoboken, New Jersey.

Published simultaneously in Canada.



*Library of Congress Cataloging-in-Publication Data has been applied for.*

Hardback: 9781394216734
ePDF: 9781394216741
epub: 9781394216758
oBook: 9781394216765

Cover Design: Wiley
Cover Image: © CasarsaGuru/Getty Images

## Contents

## List of Figures

## List of Tables

# Preface

When I was engaged in accident investigation analysis and report writing for accidents that occurred in various parts of Japan as an accident investigator for the Japan Transport Safety Board (JTSB), I had the opportunity to study scientific accident investigation analysis methods, including practical skills, at Cranfield University in the United Kingdom. What I learned and acquired there were accident mechanisms and universal methods of accident investigation and analysis and so forth, and I recognized that accident investigation and analysis is a comprehensive science involving many disciplines, such as engineering, psychology, physiology, sociology, and management. After retiring from the JTSB, I worked as the Emergency Response Coordinator and Occupational Health and Safety Section (OHS) Manager at the Okinawa Institute of Science and Technology Graduate University (OIST) in Okinawa, Japan, where I was responsible for the environmental, health, and safety and the development and operation of business continuity management (BCM).

The university experience allowed me to learn about the different approaches to safety used in many other industries, such as the maritime and service industries that I had experienced. This reality was further supported by my experience in academic exchanges, scientific papers, and literature. After OIST, as a professor (later special professor) in the Crisis Management Office at Kyushu University, I was responsible for formulating and operating the BCM against natural disasters and major accidents, as well as engaging in laboratory safety research.

At the time, I had just read a scientific paper *Nature Chemistry* describing the many accidents in university chemistry laboratories that had resulted in the casualties of faculty members, researchers, postgraduate students, and undergraduates, and that there had been no paradigm shift in laboratory safety, which never left my mind.

Scientific accident investigation analysis was not recognized within the university, nor were opportunities provided for education and training. Consequently, scientific accident investigation remained unnoticed. To prevent accidents in universities and academic institutions, I believed it was necessary for faculty members, researchers, and staff in charge of environmental safety and health office to first recognize scientific accident investigation analysis methods and the techniques for preventing accidents utilizing data derived from these analyses. It was during this time that I received an e-mail from John Wiley & Sons, Inc to write a book on accident prevention in universities and academic institutions.

This book prioritizes a clear description of theories and approaches related to accident occurrence and prevention, utilizing numerous figures, tables, and real-case examples to enhance understanding throughout each chapter. The primary audience for this book includes occupational health and safety professionals, laboratory technicians, chemical specialists, faculty members, fieldwork specialists, and diving officers. Secondary audiences encompass advanced undergraduates, graduate students, and researchers. Those who wish to learn about the mechanisms of accidents and incidents within academic and research institutions, individuals seeking to establish multiple layers of defenses to prevent errors from directly leading to serious accidents during laboratory experiments or fieldwork research activities, and those who aspire to reduce the number of accidents and incidents will find this book valuable as a text on accident theory and accident prevention. The case studies presented are all based on real accidents and incidents, and readers can derive lessons from these examples to contribute to the safety of their respective universities or research institutions. Furthermore, the accumulation of accident and incident data using the risk management and quality management process (RMQMP) approach model and other accident models introduced in this book is believed to contribute to the development of next-generation safety assurance tools for professionals, researchers, and technicians.

As discussed in Chapters 1 and 6, accident investigation reports need to be made publicly available so that many faculty members, staff, and students around the world can learn lessons from accidents and prevent accidents before they happen. The methodology is described in detail in Chapter 1, Section 1.12, Challenges in Preventing Accidents in Universities, etc. Throughout this book, common nouns are used to describe universities, laboratories, faculty members, researchers, students, etc., without using university names, research names, or individual names.

From Chapter 2s to 7, this book systematically explains the mechanisms of accident occurrence, various theories of accident prevention, and methods of accident investigation analysis in a scientific manner. Chapter 1 discusses the historical background of accident investigation alongside the current status of accident investigations in academic institutions like universities, comparing them with those conducted in other industries, and discusses the challenges universities and academic institutions face. Chapter 2 discusses useful concepts for researchers to prevent accidents, including hazard and accident concepts, the hierarchy of control for risk reduction, layers of defense to prevent accidents, the difference between near misses and accidents, and examples of lessons learned and their applications from scientific accident investigation results. Chapter 3 addresses significant accident models necessary for understanding accident mechanisms, discussing the historical development of these models, the overview and challenges of each model, and models applicable to various industries including academic institutions. Chapter 4 covers the risk management process and the safety management system (SMS) necessary to prevent accidents and mitigate damages in universities and academic institutions settings. Chapter 5 discusses human factors necessary for understanding the underlying factors of accidents. Chapter 6 provides a detailed description of the accident investigation process targeting serious accidents such as fire, explosions, casualties, and near misses that could potentially lead to serious accidents. It includes the safety of accident investigation personnel at accident sites, preservation and collection of evidence, evaluation of evidence, analysis methods, and preparation of accident investigation reports. Chapter 7 analyzes accidents involving chemical substances in laboratories, accidents during explosive usage, accidents involving high-pressure gases, accidents caused by refrigerants, accidents involving laboratory equipment, diving accidents during fieldwork activities, river accidents, mountain accidents, etc., using the RMQMP model to explain the contributing factors of each accident. Chapter 8 explains approaches to applying the BCM and SMS to universities and academic institutions as a systematic accident prevention system involving the entire organization in response to natural disasters and accidents. It also discusses preventing the recurrence of serious accidents and proactively preventing accidents from occurring.

## About the Author

Dr. Koji Fukuoka obtained his PhD (Maritime Science and Technology) from Kobe University. After serving as a Judicial Police Officer and Coast Guard Officer in the Japan Coast Guard, he worked as a Security Manager at the Mandarin Oriental, Tokyo, where he formulated business continuity management (BCM), conducted education and training and conducted workplace inspections, and was appointed as a member of the Crisis Management Committee for safety and occupational health, ensuring the safety and security of the hotel. At the Japan Transport Safety Board (JTSB) of the Ministry of Land, Infrastructure, Transport, and Tourism, he studied scientific accident investigation and analysis methods at Cranfield University in the United Kingdom, and as a Supervisory Regional Accident Investigator, he conducted numerous accident investigations and analyses and prepared accident investigation reports. At Okinawa Institute of Science and Technology Graduate University, as Emergency Response Coordinator and Occupational Health and Safety (OHS) Section Manager, he established and operated BCM and safety management system (SMS) and worked to improve safety through field research activities and workplace inspections of laboratories. In collaboration with the Provost, he established a Risk Management Committee, and as a member of the committee, he visualized and compiled the major risks of the entire university into a risk matrix, determined the priorities of risk measures as system risk management using scientific methods, and established a system for risk control by each department and laboratory. At Kyushu University, where he was a Professor and Special Professor in the Crisis Management Office, he conducted research on accident prevention, established BCM, provided education and training for faculty members and administrative staff, set up a system of BCM workplace inspections and internal audits, and transformed the organizational structure into a resilient one.

Dr. Fukuoka has been a researcher and practitioner in the fields of safety, accident prevention, and loss control in industry and academia. During his tenure at the JTSB, he served as a consultant of the International Maritime Organization (IMO) for the "National Training Course on Marine Casualty and Incident Investigation" in Manila, Philippines. He provided English-language training to government officials. Additionally, he was appointed as a Correspondence Group Analyst for III (IMO's Subcommittee on Implementation of IMO Instruments), analyzing accident investigation reports published by various countries and submitting analysis reports to the IMO. Dr. Fukuoka's notable publications include *Safer Seas: Systematic Accident Prevention* (CRC Press Taylor & Francis Group 2019) and the scientific article "A new approach for explosion accident prevention in chemical research laboratories at universities" (Scientific Reports 2022).

Dr. Fukuoka is currently President of Business Continuity Plan (BCP) and BCM Consulting, where he works as a consultant on BCM and SMS as well as accident prevention in Tokyo, Japan.

## Acknowledgments

I am indebted to many people who helped and encouraged me to complete this book; without them, this book would not have been possible. Although I cannot name all of them, I would like to thank those people whose assistance was indispensable. President Peter Gruss and Provost Mary Collins for their support in establishing the SMS and the business continuity management (BCM) during my tenure at Okinawa Institute of Science and Technology Graduate University (OIST) and the staff of the Occupational Health and Safety (OHS) who worked with me through many challenges. Top management of Kyushu University for their support in the establishment and operation of the BCM and the work related to laboratory safety. Faculty members who supported my accident prevention activities at the universities: Associate Professor Yoshinori Okada at OIST, Associate Professor Kenichi Honjoh and Associate Professor Yoshiyuki Yamagata at Kyushu University, and Professor Masao Furusho. I must express my great appreciation to Professor Graham Braithwaite, Director of Transport Systems at Cranfield University, for teaching me scientific accident investigation and analysis methods at Cranfield University in the United Kingdom.

I would like to take this opportunity to thank the many members of the Asian Conference on Safety and Education in the Laboratory who provided opportunities for discussion, debate, and presentations on laboratory safety and related topics.

# 1

# Current State of Accidents at Universities

## 1.1 Introduction

While rapid progress is being made in science and technology, many faculty members, researchers, graduate and undergraduate students have been killed in accidents involving chemical experiments in laboratories of universities and academic institutions (hereafter referred to as "universities, etc."). However, the total number of accidents that have occurred in university laboratories worldwide remains unknown, and in many cases, similar accidents have been repeated several years later at other universities. This is because time has passed without a paradigm shift or fundamental change in laboratory safety measures, and universities around the world have not made progress in collaborating to prevent accidents.

According to a questionnaire survey conducted by the author, there are few established accident investigation and analysis methods or uniform rules for accident investigation and analysis in universities, and even when accident investigation and analysis is conducted, the 4Ms (man, machine, media, and method) analysis tool mainly applied in the manufacturing industry is used. In many cases, investigation reports and lessons learned from accidents that occur within universities are never published (Fukuoka 2022a).

In addition, there are no international organizations, such as the International Maritime Organization (IMO) or the International Civil Aviation Organization (ICAO), whose objectives mainly include accident prevention. It is essential for accident investigation and analysis to determine which accident models are applicable to the industry to which it belongs and to conduct the investigation and analysis using the methods associated with those accident models. If an inappropriate accident model is used in an accident investigation, the mechanism leading to the accident will be incorrectly analyzed and the resulting safety recommendations will be misdirected in terms of preventing accident recurrence. However, many chemical accident investigations in university laboratories have not been discussed based on accident models, although case studies have been carried out. Therefore, the lack of uniform methods for accident investigation and analysis raises doubts not only about the accident factors themselves derived as a result of the accident investigation but also about the derived data results, even if the accident factor data are statistically processed (Fukuoka and Furusho 2022).

For many universities to use the lessons learned from accident analyses to reduce accidents, accident investigations must be scientific, investigation and analysis methods must be standardized, and a large amount of accident data must be collected and analyzed. The approach to this is to bring together universities, etc. around the world to formulate investigation and analysis methods and rules, including the preparation of accident investigation reports, to provide education and

*Accident Prevention and Investigation: A Systematic Guide for Professionals, Educators, Researchers, and Students*, First Edition. Koji Fukuoka.

training in investigation and analysis methods, and to establish a forum for sharing information to reduce accidents.

This book clarifies the problems that need to be solved to prevent the recurrence and prevention of accidents faced by universities, etc. in comparison with other industries, and describes approaches that can be adapted to universities, etc. First, as a prologue to the content discussed in Chapter 2 and below, issues related to accident prevention in universities, etc. obtained compared with other advanced industries are discussed.

In the subsequent chapters, the term "accident" refers to events involving injuries or fatalities occurring during educational and research activities within universities, etc., while "incident" refers to events threatening the safety of individuals other than accidents during educational and research activities, or events causing damage to facilities or equipment used during activities. Near misses are included within incidents.

## 1.2 Background

A wide variety of accidents occur in universities around the world, including explosions during chemical experiments in research laboratories. It is not possible to estimate how many accidents occur in university laboratories around the world, as the data itself are not published. The following is an excerpt from a review article by Ménard and Trant (2020) published in *Nature Chemistry*:

*Over the past ten years, there have been several high-profile accidents in academic laboratories around the world, resulting in significant injuries and fatalities ... However, the study of academic lab safety is still underdeveloped and necessary data about changes in safety attitudes and behaviours has not been gathered ... More than ten years on from Sangji's death, we can conclude that there is no evidence of sweeping, fundamental changes, nor of major paradigm shifts in how academic lab safety is approached within the discipline.*

However, in the field of ship and aircraft accidents, a social transformation, a paradigm shift, took place in the second half of the twentieth century with regard to accident prevention. This involved the following multiple factors (Fukuoka 2023):

- Shift from human error to human factors
- Development of the software, hardware, environment, liveware (SHEL) model
- Development and global adoption of safety management systems (SMSs)
- Development of accident models, such as the Swiss cheese model, which can explain the mechanisms of organizational accidents
- Shift from Criminal Investigation to Safety Investigation

The trigger for these factors was a series of serious accidents in various industries in the 1970s and 1980s that affected society as a whole. In the chemical industry, 28 employees were killed in a chemical plant explosion in Flixborough, United Kingdom, in 1974, and thousands of citizens were killed by acute poisoning in Bhopal, India, in 1984 when a chemical plant released methyl isocyanate, while the exact number of victims was never determined.

In the maritime industry, 193 passengers and crew were killed in the 1987 capsizing of the ro-ro vessel *Herald of Free Enterprise* (HFE), which was traveling to and from the Strait of Dover. In the aviation industry, in 1985, Japan Air Lines (JAL) Flight 123 crashed into Mount Osutaka, killing 520 passengers and crew and leaving only 4 survivors, making it the worst single aircraft accident. In the space industry, in 1986, the US space shuttle Challenger exploded just 73 seconds after lift-off, killing all seven crew members on board.

## 1.3 Modes and Effects of Human Error

Human error was actively studied by Norman, Rasmussen, Reason, and others in the mid to late twentieth century, and findings were developed and established as follows.

### 1.3.1 Norman's Theory

Norman (1988) emphasized that people make some errors in error-prone situations that are unrelated to specific personal characteristics. He distinguished between slips and mistakes. In other words, slips are errors of behavior, while mistakes are errors of thought.

### 1.3.2 Rasmussen's Theory

Rasmussen, Duncan, and Leplat (1987) extended Norman's theory and defined three types of performance and error. They are skill-based, rule-based, and knowledge-based. He states that people display different behavior at these three different levels.

At the skill-based level, it is an automatic method and is used when performing tasks that are routine and highly trained. The rule-based level is used when specific situations are encountered and there are already agreed rules for performing these actions. The knowledge-based level is used when a new situation is encountered that has not been experienced before. This is the highest performance level and involves people working very slowly, using all the resources imprinted in their memory through trial-and-error learning.

Rasmussen emphasizes that the process of people making decisions is not linear and that in real life people often shortcut the process. As a result, he states that the three levels outlined above can coexist at any time.

The three performance aspects of everyday life are explained, covering the process of learning to ride a bicycle. When riding a bicycle for the first time, children learn in detail from their parents and friends how to ride a bicycle, how to balance, when and how to brake, etc. When they actually ride a bicycle, they find it difficult to control it and keep going, so they try to learn to ride a bicycle through a process of repeated trial and error. This process is called knowledge-based performance. After a number of training sessions, the operating skills of riding a bicycle gradually improve and they are able to get out on the road. When they come across a traffic light, they stop, making sure that the rule is that they should not proceed at a red light. They also check that they are traveling on a particular side of the road, as they have learned from their parents. This process is called rule-based performance. They ride their bicycles every day, and after a few months, they become so familiar with it that they no longer need to pull out the bicycle maneuvering and traffic rules from their memory each time and apply them in practice. In particular, they can ride without having to think about how to maneuver it in their minds. When they encounter a red light, they are able to achieve their objectives by skipping rigid procedures experienced in the knowledge-based and rule-based performance, such as stopping reflexively. This process is called skill-based performance.

### 1.3.3 Reason's Theory

Reason (1990) expanded on Rasmussen's theory. He states that slips and lapses are behavioral errors involving skill-based performance. Mistakes and violations are errors in people's intentions, whereas mistakes occur in either rule-based or knowledge-based performance. He coined the concept of latent conditions, which was previously called latent error, and distinguished between active failure, which was previously called active error, and latent situations. Active failures are committed by operators and manifest themselves quickly, while latent conditions are caused by higher levels, such as designers and managers, and the existence of latent conditions is not exposed in the organization and lies dormant for a long time. He states that active failures are affected by latent conditions and that accidents occur due to rare combinations of these errors. This theory of his was the source for the development of the Swiss cheese model.

#### 1.3.3.1 Slips, Lapses, Mistakes, and Violations

According to Reason (1990), errors are classified into four categories. An unintended act that relates to a person's attention and an error that occurs at the stage of execution is called a slip. Errors that are unintended acts related to memory and occur at the stage of consideration are called lapses. Errors that occur because the action was intended but the plan itself was not fit for purpose due to a design error at the planning stage, and the action was carried out according to the plan, are called mistakes. If a rule exists, the act is a violation. The violation addressed here is a negligent violation without malice. For example, wearing personal protective equipment (PPE) is common knowledge and a safety rule in chemical experiments, but this applies to cases where researchers do not wear the designated PPE because they think it would be safe at their own discretion. Explained from a different perspective, the question is: "Was the action carried out a planned action?" and if the answer is "no," then an unintended action has been executed. The error in this execution is a slip or lapse. A slip is the result of not paying sufficient attention when executing an action. A lapse is a planned action not being executed due to an error in memory. "Was the action carried out a planned action?" If the answer is "yes" to the question, then the intended action was carried out and a mistake was made. The error occurred at the planning stage, which means that there was a problem in the decision-making process. Mistakes may progress more unnoticed by people than slips and lapses, and a longer period of time may pass between the execution of a mistake and its detection.

##### *1.3.3.1.1 Case Study of Slips*

- Many readers probably start their day with a cup of coffee every morning. They turn on the coffee machine, pour water, set the capsules in place, place the coffee cup in place, and press the button for their choice of blend or espresso, and their coffee of choice is ready. However, as they become accustomed to this action, doing the same thing over and over every morning, they inadvertently press the espresso button without setting the capsule in place. Only hot water is poured into the coffee cup. You realize for the first time that you did not set the capsule.
- Hotels have bathrooms, some with fixed shower heads (overhead showers) and some with portable shower heads. To use a fixed shower head, turn the left handle, shown in Figure 1.1, backward. To use the portable shower, turn the same handle toward you as for the fixed shower. The handle is marked with a "Caution! Hot Water" sign. The temperature of the hot water is adjusted using the handle on the right. The author was drenched by cold water coming out of the fixed shower head when he accidentally turned the handle because he wanted moderate hot water. Given the warning sign "Caution! Hot Water," one imagines that many people would have been drenched head-first in boiling water against their will.

Shower equipment that does not cause people to make errors would be as follows. The temperature is adjusted with the left handle and the shower head is selected with the right handle, which

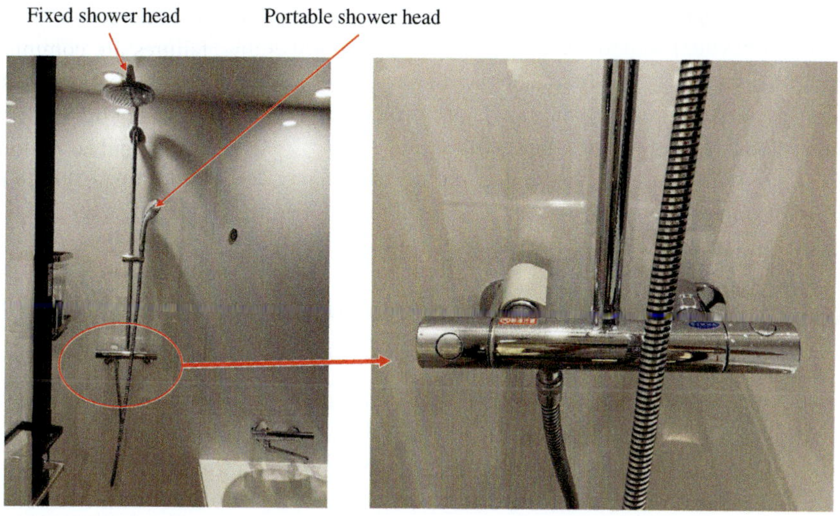


**Figure 1.1** Showerheads and their handles as a sample of slips. The left-hand handle is enlarged in the photograph on the right. The handle is marked with a red label with a warning. The large disk-shaped shower head directly above the handle is fixed. To the right of the fixed shower head is the portable shower head. These photographs were taken by the author in the same hotel in December 2022, including Figure 1.2.

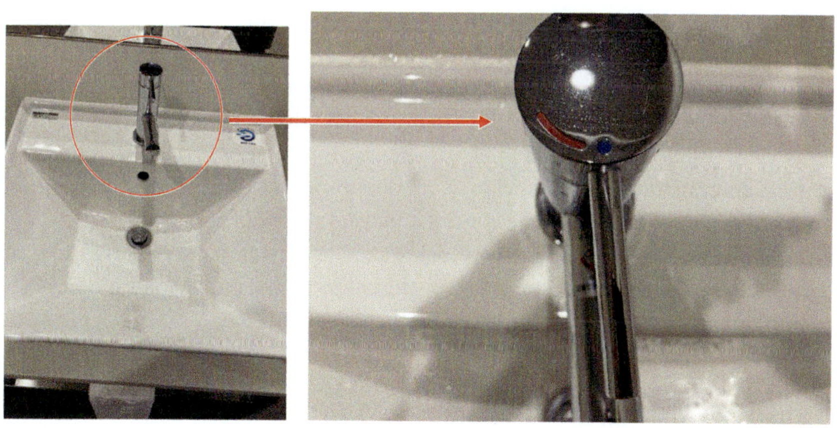

**Figure 1.2** A bathroom tap that can prevent slips. The tap in the left photograph is enlarged in the right one. The blue and red markings indicate whether water or hot water is supplied.

has a different shape. There is no room for error. This shows that human error is largely due to external factors as well as human causes.

However, the washroom in the same hotel was designed to prevent slip when using water and hot water. Figure 1.2 shows that on the tap handle, turning the handle in the blue direction provides water and turning the handle in the red direction provides hot water. One can get water or hot water without confusion. It can be seen that by matching a person's intended behavior with their thoughts, by adding shapes, signs, and diagrams to the equipment, it is possible to prevent people from making errors.

#### 1.3.3.1.2 *Case Study of a Lapse*

- You walk from your room to the kitchen with the intention of taking a teacup from the kitchen cabinet, but you enter the kitchen and stop, unable to remember what you came here to do. You go back to your room again and try to remember what you went to the kitchen to do, only to realize that it was to pick up the teacup.
- You set a single document in the photocopier to make copies and have completed the specified quantity of copies, but forget to remove the document from the photocopier. A colleague informs you that the document is still in the photocopier and you realize that you forgot to take the document out.

#### 1.3.3.1.3 *Case Study of a Mistake*

- Many people commute to work by train, bus, or other public transport to the university or workplace. When universities are on summer holidays, public transport timetables can change significantly. Without realizing this, they leave home as usual and wait at the bus stop to board their usual bus, but the bus does not come. A notice is posted on the bulletin board at the bus stop, and they learn for the first time that the bus they are using is temporarily out of service due to the university summer holidays. They rush to board another bus but are late for the scheduled work time. They realize that this error was an error in their very plan to use their usual bus to go to the university.

#### 1.3.3.1.4 *Relationship Between Errors and External Factors*

Grandjean (1969) conducted a scientific experiment on behavioral failure. He measured the frequency of errors when the arrangement of the control panel and the stove was changed. When the arrangement perceived by the person matched the arrangement of the stove and control panel, it was tested 1200 times and not a single error occurred. The three patterns with slightly different arrangements were tested 1200 times and errors occurred in the range of 76–129 times. In other words, it can be demonstrated that individuals are less prone to error when their cognitive processes align with the configuration of the control and display sequences.

The controls in the steering seats of many vehicles, such as cars, trailers, airplanes, and ships, are designed with full consideration of human characteristics, such as a person's physique, range of vision, and thinking patterns, so that errors are unlikely to occur.

### 1.3.3.2 Impact of Human Error

The examples from everyday life described in Section 1.3.3.1 show that the impact of the error only affects the person who made the error and does not cause significant harm. However, failures of caution, that is slips, have also occurred in university experiments and chemical plant operations and other industries. At a university, a timer and temperature were set on a heating device to heat a sample during an experiment, thinking that the temperature was displayed in Celsius, but the temperature was displayed in Fahrenheit instead of Celsius, causing the temperature to be set much higher than intended, resulting in the sample overheating and smoke coming from the heating device. The alarm in the laboratory was triggered and security guards responded to the incident, which did not lead to a fire.

According to the US Chemical Safety Board (CSB 2021), the accident investigation body, during the silicon emulsion production process in a chemical plant, an employee put calcium hydroxide, a pH-adjusting substance that should not have been mixed in, into a drum in which silicon hydride was stored. This generated hydrogen gas, which mixed with air, and the presence of an ignition source in the plant caused an explosion and fire, killing four employees and wrecking the plant. The drums used to store the three chemicals were of the same color and shape (55-gallon blue plastic drums), identified only by labels, and the incompatible substances were stored in close proximity. The drums were all the same shape and color, so employees could not tell them apart.

The CSB determined that the risk of the chemical combination was not defined in the US Occupational Safety and Health Administration (OSHA) standards, which was also a contributing factor in the accident. The CSB (2021) recommended that the US OSHA, the chemical policing body, review its chemical standards and even influenced legal action.

In a memory failure, lapse, during a leisure dive activity, a diver jumped in without opening the tank valve before entering the water, which made it difficult for him to breathe in the water. In the preparation before the scuba dive started, he inadvertently forgot to open the tank valve.

In an example of a vessel accident, a pilot was maneuvering a cargo ship from one port to another, when he assumed that buoy A, which marked the safe passage boundary, was buoy B, which marked the route exit, and after passing buoy A, he changed the vessel's course to the right and continued sailing, resulting in the ship grounding in shallow water. The pilot stated that he changed course as usual, thinking he had already passed buoy A and reached buoy B, which shows the exit of the passage. The accident affected many stakeholders, including the cargo ship grounded; the crew; the cargo ship's company; the Coast Guard, who maintains the buoys and enforces maritime laws and regulations; the port authority, who manages the safety of passages in the port; and the salvage company that handles the grounding ship and the property insurance company (Japan Transport Safety Board 2010).

## 1.4 Transition from Human Error to Human Factors

In the second half of the twentieth century, the increasing complexity of socio-technical systems, industrial structures, and accident causal factors made it no longer possible to apply the conventional approach to accident analysis, which is centered on human error. In other words, the idea of preventing accidents by clarifying the systemic problems that induced the error or failed to prevent the accident, rather than focusing attention on the person who caused the human error and attributing the cause of the accident, has spread, particularly in Europe and the United States. In the field of accident model, sequential accident models, such as the domino theory, which are characterized by a mechanism in which the factors of an accident appear sequentially and linearly and then accidents occur, are no longer able to clarify the actual state of accidents and all the factors behind them. The "Tenerife tragedy," an aircraft accident that occurred at a small regional airport in Tenerife Island, located off the western coast of Africa, marked the turning point for this issue and the shift in the focus of accident investigation analysis to human factors. See Chapter 3 for the development of the accident model.

### 1.4.1 The Tenerife Tragedy

#### 1.4.1.1 Summary of the Accident

In March 1977, two 747 jumbo jets of Dutch Airlines (KLM) and Pan American Airways (Pan Am) collided head-on on the runway at Los Rodeos Airport, Tenerife, Spanish Canary Islands, killing 583 passengers and crew. The accident is still the worst in aviation history. Immediately after the accident, the Air Line Pilots Association (1977), an association of pilots, published an Aircraft

Accident Report, based on interviews with those involved and evidence collected. The details of the accident based on this report are as follows, showing that a number of factors combined to cause the crash.

#### 1.4.1.2 Circumstances and Factors Leading to the Accident

Following a terrorist explosion at the terminal of Las Palmas Airport in the Canary Islands, a well-known resort destination, the airport authority temporarily closed the airport and instructed aircrafts destined to land at Los Rodeos Airport. The airport authority lifted the closure of Las Palmas Airport and resumed landings after the airport was declared safe, but in the meantime, Los Rodeos Airport was crowded with large and small aircrafts that had been instructed to change their landing sites. Los Rodeos Airport is a regional airport with one runway located at 621 m altitude, where stratocumulus and cumulus clouds frequently occur and visibility at the airport can be poor. After the reopening of Las Palmas Airport, passenger aircraft parked at Los Rodeos Airport took off one after another for the airport, but there were aircrafts parked on the taxiway.

The Air Traffic Control (ATC) first instructed KLM aircraft to proceed in the opposite direction on the runway to the takeoff preparation point. The ATC then instructed Pan Am aircraft to proceed in the same reverse direction on the runway and turn left onto the taxiway at C3, which was located near the midpoint of the runway. KLM aircraft reached the designated point at the end of the runway and awaited clearance for takeoff from ATC. Meanwhile, Pan Am aircraft was to proceed to the taxiway at point C3, but crew of Pan Am aircraft either misunderstood the instructions from ATC or overlooked C3 and continued backward down the runway toward the KLM aircraft. The Air Line Pilot Association, which investigated the accident, analyzed in its report that it was physically impossible for Pan Am, a large aircraft, to enter C3, which required a change of course at a sharp angle. It is recorded that the visibility at this time was approximately 100 m. ATC had no radar equipment installed and relied on visual observation with the naked eye to determine the situation at the airport. Both aircraft crew could not see the other, nor could ATC see the situation of the two aircrafts on the runway.

The KLM captain communicated with ATC and attempted to throttle up for takeoff, assuming that the message from ATC telling him how to fly after takeoff was the answer to the takeoff clearance. At this point, the crew's flight engineer informed the captain that he did not have takeoff clearance, but the captain prepared the aircraft for takeoff. The flight engineer did not firmly reiterate to the captain that he did not have clearance for takeoff. KLM aircraft had a team of three: the captain, the copilot, and the flight engineer. The personal relationship was such that there was an authority gradient between the three, as the captain was a person of authority at KLM and also a training instructor. The authority gradient hinders the sharing of useful safety information within the team; the captain may have been in a hurry to takeoff, as KLM aircraft planned to fly back to the Netherlands after arriving at Las Palmas Airport and could have exceeded the working time limit under Dutch working rules if he had not taken off early at Los Rodeos Airport.

Radio communication between the ATC and KLM or Pan Am aircrafts was poor, with interference during conversations, making some conversations inaudible. The ATC is a native Spanish speaker and his English was mixed with Spanish, making some parts difficult to understand. At the time of the accident, the aviation terminology used by air traffic controllers and crews was not standardized worldwide, and misleading and ambiguous terminology was used in communication. As a result, there was a mixture of assumptions.

The KLM aircraft then began its run on runway for takeoff. Shortly before the collision, the Pan Am, which was turning left on the runway toward C4 next to C3, suddenly appeared in the KLM captain's field of vision and he raised the nose of the aircraft in an emergency takeoff attempt. However, the lower fuselage of the KLM aircraft collided with the upper fuselage of the Pan Am

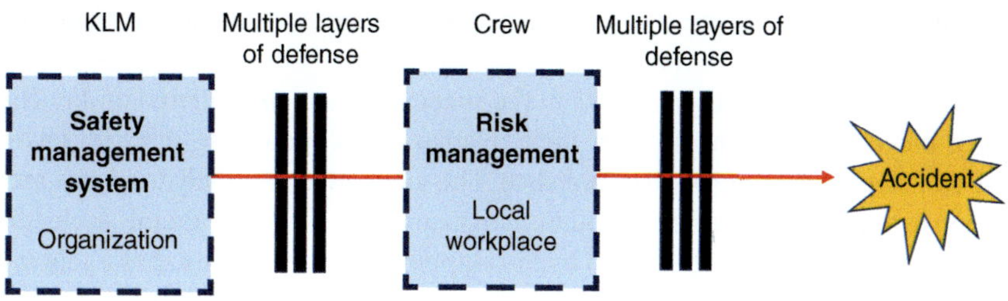


**Figure 1.3** The mechanism of the Tenerife tragedy. At the time, KLM's SMS and crew risk management were not functioning and are shown as dashed lines. The accident trajectory is indicated by the red arrow. *Source:* Adapted from Fukuoka (2024).

aircraft, causing the KLM aircraft to crash and burn and most of the Pan Am aircraft to be destroyed and set on fire; only some of the Pan Am passengers and crew survived. The contributing factors that caused the accident are listed in the following:

- Change of destination (S)
- Inconsistent (ambiguous) aviation terminology (S)
- Radio communication status between crew and ATC (H)
- Radar not installed (H)
- Narrow taxiways (H)
- Weather conditions at high-altitude airports and reduced visibility (E)
- Authority gradients, availability of time, assumptions, and low flight time of KLM captain (Lc)
- Communication between crew (Lc)
- Communication between crew and ATC (Lp)
- Language used by crew and ATC (Lp)
- Verbal guidance given to aircrafts by ATC (Lp)

Los Rodeos airport is located at a high altitude, which makes it vulnerable to weather conditions, and at the time of the accident, a new airport was under construction at a lower altitude. The Spanish and Dutch Accident Investigation Authorities have also published accident investigation reports on the accidents that occurred in Tenerife. The details of these reports are not included in this section, as this section discusses the mechanism of the accident, in particular the relationship between human factors and the occurrence of the accident.

The SHEL model indicates that the accident was not the result of a linear sequence of accident factors, as proposed by the domino theory. Rather, it suggests that a number of factors were present prior to the accident. Furthermore, the model indicates that the KLM captain's error, which was a mistake, triggered a number of latent conditions to emerge simultaneously, resulting in the accident. Figure 1.3 is a model of the accident mechanism on the KLM side. At the time of accident, KLM's SMS was not yet in place and crew risk management was not functioning. The accident trajectory shows that the accident occurred after all the multiple layers of defense had been penetrated.

## 1.5 Development of the SHEL Model

Five years before the Los Rodeos Airport Jumbo Jet accident, in 1972, Professor Edwards developed and published the SHEL model, which included social components related to the accident. SHEL stands for software, hardware, environment, and liveware. Liveware, representing human elements, includes Central liveware (Lc) and Peripheral liveware (Lp). The SHEL model was modified by Hawkins (1987).

The SHEL model has a total of five blocks in the shape of a cushion centered on Lc. The jagged edges of each block do not coincide with their relative edges, indicating that the relationships between the two elements are not well established. The model was originally used as a method for the comprehensive collection of evidence on human factors and others in aircraft accident investigations. The model has since been adopted by the shipping industry and used as a tool to prevent omissions in the collection of evidence during the accident investigation phase. Since the late 1990s, the model, together with the Swiss cheese model, has been adopted by the ICAO and the IMO, two specialized UN agencies, and has been included in manuals for ship and aircraft accident investigation and is still used in accident investigation.

The SHEL model differs from human error in that it focuses on the errors and mishaps of the person who caused the accident (hereafter referred to as the "operator") and includes all factors related to the accident that caused the human error. This includes not only the company to which the operator belongs but also regulatory authorities, inspection bodies, relevant private bodies, international organizations, etc. and all relevant social components involved in the accident can be identified. By viewing the accident as an event that occurred in the context of the interrelationship of all these social components, the clear picture of the underlying factors can be visualized.

### 1.5.1 Elements of the SHEL Model

The following is a brief description of each element of the SHEL model, adapted to accidents occurring at universities, etc. Details of the SHEL model are given in Chapter 5.

- Software S is employed by faculty, researchers, students, research groups, and fieldwork research teams at the time of an accident at the accident site. It encompasses a range of materials, including procedures, protocols, checklists, experimental plans, dive plans, climbing plans, maps, and instructions on the use of PPE and emergency response plans.
- The letter H refers to the hardware associated with the accident. This encompasses a range of equipment, including laboratory apparatus, gas storage tanks, high-pressure cylinders, ancillary equipment such as piping, laboratory equipment such as centrifuges and beakers, and other types of laboratory apparatus. It also includes diving equipment used for scuba diving and climbing equipment used for mountaineering.
- E represents the environment at the time of the accident. It encompasses specific gas concentrations that have been known to cause explosions and fires, the presence of ignition sources such as electronic devices and equipment power supplies, oxygen concentrations that can lead to oxygen deficiency, gas concentrations of specific substances that can lead to gas poisoning, weather conditions, the speed of tidal and ocean currents, seawater clarity, seabed topography, mountain topography, especially streams in the path of avalanches, marine life, plants, and animals that can harm people.
- Lc represents the condition of the operator at the time of the accident, encompassing a multitude of factors, including physical, physiological, pathological, psychological, workload management, knowledge, skills, education, and training. It encompasses physical limitations, such as instances where the operator's physique is incompatible with the facility or equipment or the underwater mask does not fit the face; sensory limitations, such as an excess of information from the senses, including sight and hearing, that the operator is unable to process; and visual limitations, such as multiple containers of the same color and shape that cannot be distinguished from each other. Other factors that may influence the outcome of the

experiment include physiological factors such as fatigue due to a lack of sleep, psychological factors such as assumptions and stress, knowledge of handling hazardous substances, including chemicals, and risk assessment methods for substances handled in experiments. Attendance at training courses on the handling of specific substances, organized by the organization, may also be a factor. Finally, the experience of such experiments and research may also influence the outcome.

- Lp includes individuals who collaborated with the operator, the organization to which they belonged, and external agencies that were involved in the accident. The presence or absence of safety-related communication with colleagues, the clarity of the research group's responsibilities and authority, the circumstances and actions taken when a near miss occurred, and the operation of the SMS are all factors that must be considered. A more detailed discussion of SMSs can be found in Chapter 4.

## 1.6 Change in the Social System and the Shift from Criminal to Safety Investigations

Twelve years after the Tenerife tragedy, a major turning point in the maritime industry occurred with the capsizing of the ro-ro vessel HFE, which was traveling between the Strait of Dover. Prior to this accident, the system for investigating marine casualties had been an ad hoc one, with a national authority setting up an accident investigation board, appointing board members, and initiating an accident investigation whenever a marine casualty occurred. The *Titanic* disaster of 1912 also led to the establishment of temporary Accident Investigation Boards in the United Kingdom and the United States, which interviewed passengers and others related to the accident, analyzed the accident, and published recommendations for safety measures before the boards were disbanded.

Following the HFE accident, the Marine Accident Investigation Branch (MAIB) was established in June 1989 as a permanent marine accident investigation body outside the Department of Transport in the United Kingdom. The trend toward the establishment of permanent marine accident investigation bodies has spread throughout the world, with some countries, such as the United Kingdom, now establishing stand-alone marine accident investigation bodies, while others have established transport safety bodies that integrate aircraft accidents with marine and rail accidents. Aircraft, marine, and rail accident investigation reports published by these accident investigation bodies use scientific accident investigation and analysis methods, such as the Swiss cheese model and the SHEL model. Two versions of the report are available: a comprehensive version comprising 20 pages or more, which is typically dependent on the severity of the accident, and a condensed version comprising two to three A4 pages. The reports are accessible to the general public via the website of the accident investigation organization, where they can be viewed and studied. These efforts have reduced the accident rate.

The reason behind this shift was that law enforcement had been focusing on what constitutes a crime and punishing people to improve law-abiding behavior and prevent recidivism, but the number of safety-related accidents had not decreased. The concept of safety investigations was not to punish people, but to establish a system of scientific investigation and analysis to identify all accident factors and issue safety recommendations to those involved to prevent recurrence, thereby reducing the number of accidents and other incidents. Although part of the industry, there is a different way of thinking and operating in the world than the theory of accident prevention through law enforcement.

### 1.6.1 HFE Capsize Accident

#### 1.6.1.1 Overview of the Accident

The HFE capsize accident is as follows (Department of Transport 1987). In March 1987, HFE left port of Zeebrugge, Belgium, bound for port of Dover, UK, with the bow doors open, and four minutes after passing the outer breakwater in port of Zeebrugge, sea water entered the engine room and other areas of the ship through the bow doors, causing the ship to capsize and kill 193 of the 539 passengers and crew on board.

#### 1.6.1.2 Circumstances and Factors Leading to the Accident

With regard to the roles of the port entry and exit operations, the role of closing the bow door was that of the assistant bosun, the bosun was the team leader who brought all the deckhands together, the first officer was responsible for the port entry and exit operations, and the master being responsible for the safe navigation of the vessel. The bow door is the door through which vehicles enter the ship's car deck when in port, and the bow door is opened to allow vehicles to be loaded on board and the bow door is closed when all vehicles have been loaded.

After all vehicles had been loaded on board at port of Zeebrugge, the first officer assumed that the assistant bosun would close the bow door and proceeded to navigate the vessel on the bridge without checking that it was closed. According to the rules of the HFE management company, the first officer was required to be on the bridge at least 15 minutes before departure to assist the master in navigating the vessel.

The assistant bosun was taking a nap in his bedroom, having finished cleaning the deck after entering port. As the leader of the deck crew, the bosun did not lock up as it was not his duty to do so. There was also no roll call to ensure that all crew members responsible for the work were present at the time of arrival and departure from the port.

Before the accident, the master had asked the company to install indicator lights on the bridge to show the open/closed status of the bow doors, but the request was not accepted by the top management. HFE has a sister ship of the same type operating at the same time, which had five previous incidents of sailing with the bow doors open. Some crew members on the sister ship claimed that the ship did not capsize when sailing with the bow doors open.

At the time, ro-ro vessels did not have watertight bulkheads, and when seawater entered the vessel, the hull structure was prone to capsizing due to the seawater trapped inside.

After leaving the port, the HFE traveled in shallow waters at 19 knots (approximately 33 km $h^{-1}$), which is a high speed for a vessel, causing squat effect, which leads ship's hull to sink and seawater to easily enter the vessel through the bow door that remained open. After the influx of seawater, the surface effect of the seawater caused the hull to lose its balance and to overturn on the shoal.

The ferry service between Belgium and the United Kingdom was highly competitive and the management put profit before safety and did not accept the master's request to repair the bilge pumps and install indicator lights. In this accident, the assistant bosun's failure to close the bow door was the direct cause of the accident, and it is not believed that he took a nap with the intention of sabotaging the work, which is considered a failure of planning.

The factors involved in this accident are listed as follows:

- Overload of the first officer during departure planning (S)
- Navigation with bow door open (S)
- Bow door closed, indicator light not installed, and bilge pump not repaired (H)
- Underwater bulkheads not fitted (H)
- Hull sinking due to squat effect in shallow water (E)
- Failure to check facts due to assumptions (Lc)

- Breakdown in teamwork and responsibility (Lp)
- Breakdown in communication between crew (Lp)
- Intense competition between companies on the Strait of Dover route (Lp)
- Policy of prioritizing profit over safety (Lp)

## 1.7 Widespread Use of SMSs

The SMS is based on the quality management system originally created by the manufacturing industry to improve products and services. It has been modified to ensure safety when aircraft and ships are in operation. The concept of a quality management system was originated in the United States, but after the Second World War, the Japanese manufacturing industry recognized the importance of quality management systems and developed its own Total Quality Management (TQM), a method of pursuing quality improvement by considering quality as part of business management. As Japan's manufacturing industry developed, the Japanese version of TQM spread to the United States and the rest of the world, and the ideas and theories of TQM were reflected in the International Organization for Standardization (ISO) quality management system, which became an international standard as ISO 9001 (2008).

The SMS is based on the ISO 9001 concept, which was adopted by the IMO in the late twentieth and early twenty-first century as the International Safety Management Code (ISM Code) and made it mandatory worldwide to establish an SMS in accordance with the Code. The purpose of the ISM Code is to ensure safety and pollution prevention through the implementation of SMS, mainly for ships and ship management companies. This has led to the implementation of SMS on all ocean-going vessels and shipping companies around the world.

ICAO has contributed to the dissemination and development of SMS by establishing Annex 19 Safety Management and the ICAO Safety Management Manual. This trend can also be seen in other industries, and in the chemical industry, an understanding of SMS is essential when reading between the lines of the CSB accident investigation reports.

The specific interaction between the central liveware and the peripheral liveware in the SHEL model shows how organizations far from the accident site are involved in the accident, which is simply represented by the industrial SMSs listed in Figure 1.4. The aviation industry refers to these as Safety Risk Management and Safety Assurance (Stolzer, Halford, and Goglia 2008), and similar systems exist in other industries such as maritime industry.

For universities, etc., SMSs are important from an accident prevention perspective, but it appears that few universities, etc. have implemented the same systems as an organization. The CSB, the US-based accident investigation organization that usually investigates accidents in the chemical industry, but has investigated accidents that have occurred at universities, has closely examined each process of the SMS and analyzed whether there is a discrepancy between the actual accident-contributing factors and the content of the SMS. The SMS is also necessary to understand the mechanism of accidents that occurred at the university and the chemical plant, which are included in this book as real-life examples.

### 1.7.1 SMSs in the Industry

Figure 1.4 shows a simplified SMS implemented in organizations (e.g. companies) in the transport industry. The left side of the figure shows safety risk management with the risk assessment process as the core. Hazard identification, risk analysis, risk estimation, risk evaluation, and risk treatment are carried out in turn, and if the risk associated with the action is sufficiently low and acceptable

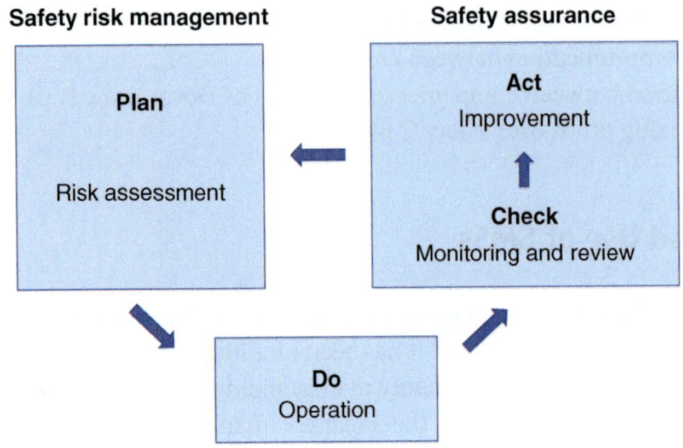


**Figure 1.4** Safety management systems.

to the organization, the aircraft, ship, or other means of transport can be operated. If the risk is unacceptable to the organization, the risk must be reduced, and the same process is repeated to reach an acceptable level. The right-hand side is called safety assurance and indicates that the operation is monitored following a risk assessment. If new risks arise due to changes in the environment or other factors, or if the risk control measures taken are no longer effective, the operation is stopped, and the risk assessment on the left-hand side is repeated to minimize the risk and the operation is restarted.

When an accident or incident occurs, the organization conducts an accident investigation. When the investigation and analysis reveal that there is a problem with the system, the defect is rectified. This ensures that the system is optimally designed and operated.

These two processes are combined into a single process known as SMS, which involves conducting risk assessments, continuous monitoring, internal and external audits of the system, investigating accidents and incidents when they occur, and correcting deficiencies identified through monitoring, auditing, and investigations, thereby continuously improving safety regardless of the fluctuating environment. Organizations are required not only to carry out risk assessments as a matter of course but also to investigate and analyze accidents and incidents and ensure that problems are rectified. For further details on risk management and SMS, please refer to Chapter 4.

A similar standard is ISO 14001:2015 (2015) or earlier OHSAS 18001:2007 (2007), which also has a quality management system, or ISO 9001, at its core and is almost the same in concept to Figure 1.4.

## 1.8 Development of the Swiss Cheese Model

During the 1970s and 1980s, a number of complex and major accidents occurred and the Swiss cheese model was developed to be applied to these accidents. The model was developed to fit the complex and large accidents described in Section 1.4. According to Reason (1997), accidents are generally caused by a combination of latent conditions and active failures, but can also occur without active failures due to hardware problems. Latent conditions are deficiencies in a number of preexisting layers of defenses; active failures are errors or mistakes made by operators in the field; and active failures themselves are also caused by latent conditions and often trigger accidents. Sometimes, as in the case of the Space Shuttle Challenger accident, it is not the operator's error that causes the accident.

A hazard does not immediately lead to an accident, but there are layers of defenses (e.g. SMSs, risk assessment) between the hazard and the accident, and as long as these layers are functioning, accidents do not occur.

Latent conditions are revealed during incidents and accidents through comprehensive and scientific investigation and analysis. These latent conditions exist before the accident, but their existence is not recognized by the operator, team members, or the organization. The time period during which latent conditions are generated can be diverse, such as when the designer of the system installed the system or when factors occur among employees who did not follow procedures when they were using the system before the accident.

In the field of aircraft accident investigation, the use of the SHEL model to collect a comprehensive range of evidence scattered around accident sites, combined with the Swiss cheese model theory that accidents are caused by many latent conditions, has reduced the aircraft accident rate, although this also includes the effect of follow-up actions with the parties involved after the accident investigation report has been published.

## 1.9 Characteristics of Universities and Other Academic Institutions and SMSs

SMSs are not widely used in universities and other academic institutions. However, it is possible to operate risk management processes and SMSs calibrated from those used in industry and applicable to universities and other academic institutions, details of which are given in Chapter 4. This section describes the characteristics of universities as opposed to industry.

The CSB (2021) describes the organizational structure of the university in its accident investigation report as follows. In universities, the principal investigator (PI) has significant authority over his or her own research. A so-called academic "feudal system" has been formed, whereby departments are divided into smaller units, each of which has a person in charge, i.e. a faculty member. These persons in charge are nominally subordinate to a higher person, but in practice they act almost as they wish, as long as they do not interfere excessively in other feudal domains. Within the organizational structure, the President, Provost, and Vice President of research can exert authority over PIs, while the chair of the department has difficulty influencing all faculty members and PIs in the department, as he or she is also responsible for PI duties.

In industry, however, there is a thorough top-down and line management approach in many companies, with the CEO issuing health and safety commitments, the health and safety department preparing a detailed health and safety plan based on the commitments, which is then rolled out horizontally in each department and communicated to employees at the end of the line, with the plan then being implemented in the local workplace. In this way, compliance with the health and safety rules and guidelines developed by the Health and Safety Department is ensured throughout the organization, and any deficiencies identified during workplace inspections are followed up by the Health and Safety Department until they are rectified in the workplace concerned.

Regarding the use of chemicals, OSHA (2013) in the United States outlines the differences between how chemicals are used in laboratories and in industry in the following. In industry, workers handle chemicals continuously or repeatedly in standardized processes and are exposed to large amounts of hazardous chemicals in the environment. However, researchers in laboratories at universities, etc. use a greater variety of hazardous chemicals than in industrial workplaces, but they use fewer hazardous chemicals at any one time, and their chemical handling procedures may change frequently.

According to the CSB (2010), each PI's work with chemicals can be very different, and as a result, research hazards can vary significantly from laboratory to laboratory. Under one PI, researchers or students in the same laboratory are not working on one project, but on different projects that may pose diverse safety hazards to them. In its conclusion, the CSB states that industry guidelines may not be suitable for direct application to the dynamic environment of university laboratories without modification. Therefore, taking into account the differences in the organizational structure and operation of chemicals in universities, it is necessary to modify the industrial SMS so that it can be used by universities, etc. to establish a SMS to ensure the safety of universities, etc.

## 1.10 Explosion Accidents in Universities

Accidents at universities are generally not published. If they are published, it is a short report of about one A4 page with a summary of the accident and the main causes of the accident. However, in the United States, the CSB publishes the full version of university accidents, but this is a rare case: the CSB is a federal agency established in 1990 under the Clean Air Act and deals exclusively with chemical accidents that occur in the chemical industry, such as chemical plants, rather than accidents at universities and other academic institutions.

### 1.10.1 Reanalysis of Explosion Accidents at Universities

The following is a reanalysis of the accident investigation reports published by the CSB using the risk management and quality management process (RMQMP) approach model (Fukuoka 2019), developed by the authors as described in the following (Fukuoka and Furusho 2022).

The accident occurred in January 2010 when a postgraduate student at University A was experimenting with a scale-up of the chemical nickel hydrazine perchlorate (NHP); the recommended use of NHP was 100 mg, but this was exceeded and the powder exploded while the NHP was being stirred with a pestle, injuring the postgraduate student. Measures to prevent the recurrence of this accident are discussed in Chapter 2, the application of the accident model in Chapter 3, the SMS in Chapter 4, and safety recommendations in Chapter 6.

#### 1.10.1.1 Local Workplace Procedures at the Time of the Accident

There were no written procedures, etc., and dates and amounts of reactants were not documented in the experimental records.

##### *1.10.1.1.1 Use of PPE*

1. It was an individual's choice whether or not to wear goggles, and the wearing of PPE depended on whether or not the individual perceived the experiment to be dangerous.
2. Several postgraduate students used regular power glasses for eye protection.

##### *1.10.1.1.2 Psychological Factor Due to Familiarity, Complacency, and Assumptions*

The postgraduate students had to consult the PI before making a decision to scale up. However, two postgraduate students had found in their pre-accident experience that small amounts of the compound did not ignite or explode on impact when wet with water or hexane and assumed that the hazards would be controlled in a similar way even with large amounts of NHP. Postgraduate students had reduced risk awareness of NHP due to previous experience and familiarity.

#### 1.10.1.1.3 *Knowledge, Skills, Education, and Training such as Conducting Risk Assessment, Including Hazard Identification*

1. The postgraduate student who synthesized the NHP had only completed a prior literature review to familiarize himself with similar high-energy compounds and had not received any research-specific training.
2. The PI or faculty did not assess the postgraduate student's understanding of the risks prior to starting research on high-energy materials.
3. Most chemistry students, including the injured postgraduate student, had not taken the general laboratory safety training offered by Environmental, Health and Safety (EHS) staff online and in person.

### 1.10.1.2 Communication

1. Safety restrictions, such as the 100 mg limit on the amount of compounds that could be synthesized, were only communicated verbally to some postgraduate students by the two PIs in the same laboratory and were not shared with everyone.
2. In the laboratory, there were two communication tools for sharing information about daily research activities: group meetings and laboratory notebooks. Before the accident, weekly group meetings were held between the postgraduate students and the PIs, but the focus was mainly on the results of the experiments and not on the actual research activities or work safety.

### 1.10.1.3 Accidents and Near-miss Response

Prior to the January 2010 explosion, two previous near misses had occurred within the same research group. The near miss occurred during scale-up when a postgraduate student inadvertently got the units of measurement wrong and produced an excess of a known high-energy material. The PI did not investigate the cause of the excessively produced known high-energy material.

### 1.10.1.4 Development and Operation of General SOPs and Experiment-specific SOPs

Experiment-specific SOPs required for the synthesis of NHP and other high-energy substances had not been developed. The Chemical Hygiene Plan (CHP) developed in 1997 did not address physical hazards, and the CHP was developed in accordance with OSHA's Occupational Exposure to Hazardous Chemicals in Laboratories Standard (29 CFR 1910.1450).

### 1.10.1.5 Approval of Changes to Experimental Plan

The postgraduate students in the laboratory did not need to obtain permission or seek approval from the PI before changing variables in the research experiment, including increasing the amount of the synthesized substance or changing the temperature.

### 1.10.1.6 Systematic Training and Education Programs

There was a lack of a systematic education and training program in line with the university's safety management: (1) chemistry postgraduate students were not required to attend general laboratory safety education and training, and there was no record of attendance for this education and training. (2) The safety education and training received by most postgraduate students in the Department of Chemistry consisted of learning from safety video aimed at undergraduate students, which is provided by the American Chemical Society. The video did not explain the need for preexperimental risk assessment.

#### 1.10.1.7 Workplace Inspections, Audits, and Corrective Actions

Both the safety advisor and the EHS safety inspector conducted safety audits and inspections of 118 chemistry laboratories on campus, many of which failed to take corrective action. In academic institutions, PIs may view laboratory inspections by external agencies as a violation of academic freedom.

#### 1.10.1.8 System for Conducting Accident and Near-miss Investigations

Although near misses occurred during scale-up prior to the accident, they were not investigated by the university, and near-miss information was not shared outside the laboratory.

#### 1.10.1.9 Coordination Between EHS and the Laboratory

Safety inspections of laboratories under the responsibility of EHS were carried out in the absence of the PIs and were not coordinated with the laboratories. Safety inspections were not seen as an opportunity to ascertain the reality of how the laboratory is ensuring safety.

#### 1.10.1.10 Organizational Structure to Ensure Safety

(1) EHS was not under the authority of the Vice President for Research and had no authority to close the laboratory based on the results of laboratory safety inspections. (2) There was no person or organization responsible for the safety management of the university, including the development and operation of the CHP.

#### 1.10.1.11 External Organizations Relevant to Accident-Contributing Factors

(1) OSHA's laboratory standards, which University A used as a reference in the development of its CHP, dealt with hazardous chemicals that could cause acute or chronic adverse health effects and did not deal with the physically hazardous chemicals relevant to this incident. (2) The Department of Homeland Security (DHS), the research funding agency, did not have safety regulations specific to the research that was being conducted at University A, missing an opportunity to influence the safety of research at the university. (3) In industry, various methodologies are used to assess hazards with regard to workplace and laboratory safety, and guidelines and examples have been developed and published. However, no guidelines have been published that provide various methodologies for risk management focusing on safety in university laboratories.

### 1.10.2 Summary of the University's SMS

Judging from the content of the accident-contributing factors identified for the explosion that occurred during an experiment at University A, ranging from "Accidents and near-misses response" to "Organizational structure to ensure safety," it can be seen that the SMS at the university was weak. On the left side of the SMS in Figure 1.4, the risk assessment process is not planned and implemented, and on the right side, the deficiencies identified through monitoring and review process have not been addressed.

#### 1.10.2.1 Safety and Academic Freedom

The challenge is the relationship between academic freedom and safety assurance. Workplace inspections of laboratories by the safety inspector were carried out in the absence of faculty members and PIs, resulting in a lack of coordination between both parties. Furthermore, workplace inspections themselves were a mere formality, indicating that inspections had taken place but not ensuring safety. In addition, due to the organizational structure, EHS, which conducts the workplace inspections, did not have the authority to close the laboratory if the safety inspector determined, based on factual information, that the risk in the laboratory exceeded acceptable level.

## 1.11 Consequences of Inexhaustive and Nonscientific Accident Investigation

This section discusses the Niseko Annupuri avalanche accident in Hokkaido, Japan, as an example of the importance of systematic and scientific accident investigation and analysis. This real-life example provides us with a lesson as to why people have been caught and victimized by avalanches three times in the same place over a long period of time.

Tourists and others have been killed by surface avalanches at the same location, commonly known as Spring Falls, on three occasions in 1990, 1998, and 2017. The accident in February 1998 in which two snowboarders were killed and injured was the subject of a court case and its details were published (Sapporo District Court 2000). In the 2017 accident, snow and ice research societies conducted survey of accumulated snowfall and measured the amount of new snow near the accident site after the accident occurred, but no systematic accident investigation analysis was conducted in either case.

In the 1998 accident, some contributing factors were identified in the judicial decision, as shown in Table 1.1, but there were insufficient contributing factors on the part of guides. Furthermore, the lack of comprehensive accident investigations involving the organizations managing the Niseko Annupuri ski resort, the company to which the guides belonged, and the local authorities meant that many organizational factors were not identified in the accident.

According to the judicial decision, the guides who led the snowboarders were aware that the Sapporo District Meteorological Office had issued an avalanche warning early that morning; that Spring Falls was designated as an avalanche danger zone; that they did not carry avalanche beacons, zombie poles, or probes as avalanche safety equipment in case of emergency; and that they were unable to take any action after being caught in the avalanche.

The only action taken to prevent the recurrence of avalanche accidents was the distribution of leaflets by the local safety liaison council after the 1990 avalanche accident, indicating that Spring Falls was an avalanche danger zone. Recurrence prevention measures were only taken at one point against a wide range of aspects, and lessons were not drawn from criminal investigations conducted by the police and prosecutors or court rulings. Then, 19 years later, in 2017, two skiers were caught in an avalanche at the same location, and both were killed and injured. The avalanches at this time of year are surface avalanches with high avalanche speeds and are said to be difficult to avoid if a person is in the path of the avalanche.

What this series of tragic events teaches us is that the analysis and conclusions of the inexhaustive accident investigation, and the safety measures taken as a result of those conclusions, were measures taken in the absence of accident factors that should have been present. This reality, and in the light of the theory of the Swiss cheese model, means that many layers of defenses have not been built up and that accidents are likely to recur. It demonstrates that if accident investigation and analysis are not carried out scientifically, covering all accident-contributing factors and targeting safety measures at the components of the social system related to the accident occurrence as a wide range of aspects rather than a point, accidents will recur over the years.

**Table 1.1** Analysis of avalanche accidents on Mt. Niseko Annupuri.

| SHEL element | Subdivision | Accident-contributing factors |
|---|---|---|
| Software | Local workplace procedures at the time of the accident | The originally planned destination was changed due to bad weather.<br>The two tour guides did not explain the specific route of the Spring Falls and the avalanche. |
| | Practices and regulations used in the field | NA |
| | Emergency response plans | The tour guide noticed the avalanche and shouted for all four of them to run away, but they were all caught in the avalanche.<br>One guide managed to get out of the snow on his own, while the other three were found and rescued by rescue personnel. |
| Hardware | Status of facility and equipment relevant to the accident | Avalanche beacon, Zombie pole, or probe not carried with. |
| Environment | Weather conditions | Snowfall increased by 44 cm between 9 p.m. on 24 May and 9 a.m. on 28 May 1998; Sapporo District Meteorological Observatory issued heavy snow and avalanche warning in the early morning of the accident day. |
| | Geographical features of the mountains | The rest area is a stream channel with little tree cover and a passing area in the event of an avalanche on the upper part of the stream. |
| | Features of mountain trails | NA |
| Central liveware | Physical and sensory limitations | NA |
| | Physiological factor (accumulation of fatigue) | NA |
| | Pathological factors (preexisting medical conditions) | NA |
| | Psychological factors (psychological pressure) | NA |
| | Psychological factors (familiarity, complacency, assumptions) | One tour guest knew from leaflets and other information that the area around the Spring Falls was an avalanche danger zone but was not worried because they were accompanied by guides. |
| | Workload management | NA |
| | Knowledge, skills, education, and training | The two tour guides were aware that there had been heavy snowfall for several days prior to the accident, that a heavy snow and avalanche warning had been issued that morning, and that the snowfall on the day was about 30 cm. |
| Peripheral liveware | Communication | NA |
| | Teamwork | NA |

| SHEL element | Subdivision | Accident-contributing factors |
|---|---|---|
| | Accident and near-miss response | An avalanche accident occurred at Spring Falls on 15 January 1990.<br>The Niseko Ski Resort Safety Council produced and distributed leaflets, designating Spring Falls as an avalanche danger zone. |
| | Establishment of criteria for suspension of tour guides | NA |
| | Development and operation of SOPs | NA |
| | Operational arrangements for emergency response plans | NA |
| | Organizational structure to ensure safety | NA |
| | External organizations relevant to accident-contributing factors | NA |

*Note:* NA in the table means that the accident-contributing factors based on the SHEL model were not included in the judgment text.

## 1.12 Challenges in Preventing Accidents in Universities, etc.

What is the current state of accident investigation and prevention in universities, etc? In the case of university accidents, accident investigations are rarely carried out and published for the purpose of a full report, and are usually in the form of a simplified report, in which only a summary of the accident and the main causes of the accident are highlighted. The actual situation at universities, etc. is therefore similar to the accident investigation of avalanche accidents described previously.

Scientific papers do not provide an exhaustive and scientific analysis of accident investigation, but rather case studies (Kemsley 2009), in which partial accident factors are addressed and discussed. These studies are informative with regard to safety measures for the factors addressed, but do not provide measures to prevent recurrence and fall into the category of symptomatic treatment. To use an analogy with a human disease, a person's back hurts, so a poultice is applied to the painful area, which temporarily relieves the pain. However, the pain will recur or persist unless the cause of the pain is thoroughly identified by a thorough examination and the cause is treated. The same applies to accidents: symptomatic treatment will result in the accident recurring.

There are safety standards for chemical experiments, *Guidelines for Chemical Laboratory Safety at Academic Institutions*, published by the American Chemical Society (2016). The guidelines comprise four pillars and explain these in detail: hazard recognition, risk assessment of hazards, minimizing the risk of hazards, and emergency response preparedness for uncontrolled hazards. The guidelines state that accidents and near misses should be reported and shared with people in the laboratory and the department and that accident investigation and accident reporting are important to prevent accident recurrence. However, it does not go into detail on the types of accidents and near misses that should be shared, the methods of accident investigation and analysis, the format of accident investigation reports, and how the information contained in accident investigation reports can be used to prevent accidents.

The reality and background of the accidents that occurred at many universities, as pointed out by Ménard and Trant (2020) in a review article in *Nature Chemistry*, is the same as the reality and background of the avalanche accidents that occurred repeatedly three times over the years at the Spring Falls on Mt Niseko Annupuri. In other words, at the time of the accidents and near misses, no thorough and scientific accident investigation and analysis are carried out, and all accident-contributing factors are not visualized. Therefore, even if safety measures are put in place that do not include factors that should have been present from the outset, similar accidents will still occur at the same university or elsewhere in the world. What universities, etc. need to do to overcome this situation is clear from what has been said so far in this chapter.

The CSB (2010), which published an accident investigation report on the explosion that took place at University A, stated the following:

> "In previous years, other institutions (except for CSB) have attempted to collect data on laboratory incidents, but no nationwide reporting system for tracking near misses and incidents exists; as a result, academia is missing a significant opportunity to communicate, educate, and improve laboratory safety."

Following the identification of issues in comparison with other advanced industries in accident prevention described in this chapter, the first steps to be taken to achieve the goal of preventing the recurrence and proactively prevention of accidents occurring in universities, etc. in the future will include the following items (Fukuoka 2022b).

1. First of all, the accident investigation and analysis methods of universities, etc. need to be standardized. Although there are currently numerous accident models available, it is recommended that an accident investigation analysis methodology based on the scientifically proven Swiss cheese model and the SHEL model be used. Reproducibility is one of the most important elements of a scientific paper. To be reproducible, the methodology must be scientifically proven. In addition, once the data from accident investigation reports have been accumulated, the underlying data must be collected to the same standard when processing the data to predict the circumstances under which an accident might occur on the basis of the data. Unification of accident investigation and analysis methods is therefore necessary for the future development of science and the improvement of safety in education and research at universities, etc. See Chapter 3 for accident models, including the Swiss cheese model, and Chapter 5 for details of the elements of the SHEL model.
2. Next, there is a need to develop common accident investigation and analysis guidelines for use by universities, etc. worldwide. The guidelines should describe the purpose of accident investigation, the methodology of accident investigation analysis, how to write accident investigation reports, how to follow up corrective actions, and how to share lessons learned. The guidelines will specify the content of the accident investigation report, but it is necessary to define the case for preparing a full and exhaustive accident investigation report and the case for preparing a simplified report. Furthermore, a distinction must be made between safety investigations and investigations for punitive purposes. The purpose of a safety investigation is to prevent accidents, not to punish people. This shall be clearly stated in the accident investigation report. The accident investigation personnel shall also explain the above to the persons involved in the accident at the beginning of the accident investigation.

   When writing an accident investigation report, the special features of universities, etc. must be taken into account. Universities and other academic institutions research undisclosed cutting-edge technologies for innovation. Exposure of such technology would cause significant damage and researchers would have a negative view of accident investigations and would not cooperate with accident investigators. Accident investigation reports should focus only on safety-related processes, without exposing such cutting-edge technology. The report should be anonymous, so that personal information, such as the name of the university, research institute, laboratory or individual, is not revealed. See Chapter 6 for further information.

3. Once the above two points have been addressed, the accident investigation personnel at each university, etc. must be educated and trained. This is because the quality of an accident investigation report depends to a large extent on the skills of the person in charge of conducting the accident investigation. Therefore, education and training materials shall be developed, and face-to-face training shall be provided. It is recommended that the content of the training material should include Chapters 2–7.

   This book does not provide a practical exercise for accident investigation personnel. However, this can be achieved through the use of this book. First, divide the trainees into several groups. Using one of the cases in this book, the trainers create a simulated situation of the contributing factors behind the accident, and the trainees interview the trainers and their colleagues to gather factual information. Based on this factual information, an accident analysis is carried out. The causes and contributing factors of the accident are identified and safety measures are recommended to prevent a recurrence. These processes can be practiced by representatives of each group, who make presentations and discuss them.
4. It is then recommended that an international conference of accident investigators be organized on a regular basis. The aim of these meetings is to improve the skills of accident investigation personnel and to exchange information useful for accident investigation and analysis, including the exchange of lessons learned from accident investigation results. The results of international conferences should be published on a website after the conference discussion of what should be made public, in the same way as is done at academic conferences, so that universities, etc. around the world can share information.

It should be noted that accident investigation and analysis and accident prevention measures should be considered and implemented in the context of the SMS referred to in Section 1.7, which is effective in preventing the recurrence of accidents and in proactively preventing accidents from occurring. The introduction and widespread use of SMS will be an important issue for universities, etc. together with other issues such as the introduction of scientific accident investigation methods and the visualization of all accident-contributing factors, taking into account the specificities of the organizational structure of universities, etc.

The benefits to universities, etc. of implementing the four steps mentioned above may include the following (Fukuoka 2022b):

1. First, if universities and other academic institutions around the world participate in this scheme aimed at preventing the recurrence and proactive prevention of accidents, the effect of economies of scale will make visible all accident-contributing factors that have been latent and unexposed up to now. These include, for example, fires, explosions, fatalities, injuries, and serious health hazards resulting from field research activities and chemical experiments, which can be displayed by type of accident. If these accident-contributing factors are present before the start of an experiment at the reader's university or other academic institution, the results of the analysis obtained can be used as lessons learned to prevent accidents from occurring, for example, by taking safety measures before the experiment is carried out.
2. Once a large amount of accident data have been collected, it can be analyzed to derive accident patterns for each type of accident. These patterns could include, for example, fires and explosions involving certain chemicals; explosions involving gas mixtures; marine research diving

accidents; river accidents; and research accidents in mountains, forests, wilderness, etc. New safety education and training programs can be developed and delivered using accident patterns. These need to be more practical and effective than the textbooks of the past.

3. At present, accident data from universities, etc., which is one of the aims of this book, have not been collected, so it is not possible to predict accidents using artificial intelligence based on vast amounts of credible data from the past. Imagine what would happen if we could use data showing the mechanisms of accident occurrence, when vast amounts of reliable and reproducible data have been accumulated. I believe the time will come when AI will be able to predict the occurrence of accidents and tell people what they should do to prevent accidents before the experiment starts. In other words, there will be innovation and a paradigm shift in accident prevention at universities, etc.

## 1.13 Conclusion

Numerous accidents of various types have occurred in universities, etc., and similar accidents have occurred repeatedly around the world, but the reality of these accidents has not been clarified. Looking at other industries, the second half of the twentieth century saw a paradigm shift in marine and aircraft accidents, with national accident investigation agencies carrying out scientific and thorough accident investigations and analyses, and making the lessons learnt available to the maritime and aviation industries, which were then applied by companies in these industries, contributing to a reduction in the accident rate, particularly in the case of aircraft accidents. By analyzing the situation in other industries that have preceded and succeeded in reducing accidents, the challenges of the safety environment for universities, etc. have been identified.

Accident models represent the mechanisms by which accidents occur. It is important to understand the historical background of the applicable accident models and how the models work before using them for accident prevention. The treatment of back pain has been used as an analogy to illustrate one of the problems associated with accident prevention in universities, etc. Not only explosions in chemical laboratories but also accidents in universities, etc. have a cause-and-effect link, and there are many accident-contributing factors. For these reasons, the epidemiological accident model is applied, as discussed in this chapter. In other words, to prevent accidents, universities, etc. should conduct accident investigations analysis using the concept of the Swiss cheese model, which is classified as an epidemiological accident model, and the accident investigation analysis tool based on the SHEL model, which allows a comprehensive investigation of accident-contributing factors. In addition, when adopting the concept of the Swiss cheese model and the SHEL model, it will be necessary to understand the SMS that was not in operation in many universities, etc. and to implement a modified version of this system well adapted to them.

It takes a long time for an organization to accumulate data on accidents that have occurred in the organization, and for accident prevention measures to be effective, it is essential to analyze accident-contributing factors for each type of accident and clarify accident patterns. However, if a large number of universities, etc. were to come together to formulate rules, including investigation and analysis methods and report writing, to provide education and training in investigation and analysis methods, and to establish a forum for exchanging information to reduce the number of accidents, it would be possible to reduce the number of accidents in a short time through development that takes full advantage of economies of scale.

Once we reach the stage where a large amount of reliable and reproducible data have been accumulated, and where we can use the data to show the mechanism of accident occurrence, AI will be able to inform researchers of the possibility of accident occurrence before they start a particular experiment, and then show them safety measures to prevent the accident, enabling researchers to proactively prevent accidents before they occur. In other words, the time will come when researchers will be able to use AI to prevent accidents, which will lead to innovation and a paradigm shift in accident prevention in universities, etc.

## References

Air Line Pilots Association (1977). Aircraft Accident Report, Engineering and Air Safety, Washington, DC. Pan American World Airways Boeing 747, N737PA KLM Royal Dutch Airlines Boeing 747, PH-BUF, Tenerife, Canary Islands, March 27, 1977.

American Chemical Society (2016). *Guidelines for Chemical Laboratory Safety in Academic Institutions.* Washington, DC: American Chemical Society.

Department of Transport, UK (1987). *MV Herald of Free Enterprise Report of Court no.8074 Formal Investigation.* London: H.M.S.O.

Fukuoka, K. (2019). *Safer Seas: Systematic Accident Prevention.* Boca Raton: CRC Press Taylor & Francis Group.

Fukuoka, K. (2022a). New approach to prevent explosion accident. *Kagaku* 77 (9): 31–36. (Japanese).

Fukuoka, K. (2022b). A new approach to preventing accidents occurring at universities. *Presentation Presented at the 9th Asian Conference on Safety and Education in Laboratory (ACSEL) in Singapore*, Singapore (17 November 2022).

Fukuoka, K. (2023). Accident and safety at the universities and chemical industry. *Kinka* 75 (10): 1–6. (Japanese).

Fukuoka, K. (2024). Forward-looking safety with the aim of developing scientific diving. *Lectures Presented at the Symposium to Educate about Scientific Diving in Fukuoka*, Japan (6 June 2024).

Fukuoka, K. and Furusho, M. (2022). A new approach for explosion accident prevention in chemical research laboratories at universities. *Scientific Reports* 12 (1): 3185.

Grandjean, E. (1969). *Fitting the Task to the Man: An Ergonomic Approach.* London: Taylor & Francis.

Hawkins, F.H. (1987). *Human Factors in Flight.* Aldershot: Gower Technical Press.

ISO 9001:2008 (2008). *Quality Management Systems - Requirements.* Geneva: International Organization for Standardization.

ISO 14001:2015 (2015). *Environmental Management Systems – Requirements with Guidance for Use.* Geneva: International Organization for Standardization.

Japan Transport Safety Board (2010). Marine accident investigation report: Bulk carrier *HANJIN BRISBANE* grounding. MA2010-8. (Japanese).

Kemsley, J. N. (2009). Learning from UCLA. *Chemical and Engineering News* 87 (31): 29–34.

Ménard, A.D. and Trant, J.F. (2020). A review and critique of academic lab safety research. *Nature Chemistry* 12 (1): 17–25.

Norman, D.A. (1988). *The Psychology of Everyday Things.* New York: Basic Books.

Occupational Safety and Health Administration, USA (2013). Appendix A to §1910.1450-National Research Council Recommendations Concerning Chemical Hygiene in Laboratories (Non-Mandatory).

OHSAS 18001:2007 (2007). *Occupational Health and Safety Management Systems – Requirements.* OHSAS Project Group.

Rasmussen, J., Duncan, K., and Leplat, J. (1987). *New Technology and Human Error.* New York: John Wiley & Sons Ltd.

Reason, J. (1990). *Human Error*. New York: Cambridge University Press.

Reason, J. (1997). *Managing the Risks of Organizational Accidents*. Surrey: Ashgate Publishing Limited.

Sapporo District Court (2000). Judicial decision of the Sapporo District Court, Otaru Branch: Niseko Annupuri "Haru no Taki" avalanche disaster case. (Japanese).

Stolzer, A.J., Halford, C.D., and Goglia, J.J. (2008). *Safety Management Systems in Aviation*. Aldershot: Ashgate Publishing Limited.

US Chemical Safety Board (2010). Texas Tech University laboratory explosion, No. 2010-05-l-TX. https://www.csb.gov/texas-tech-university-chemistry-lab-explosion/ (accessed 10 May 2023).

US Chemical Safety Board (2021). Chemical reaction, hydrogen release, explosion, and fire at AB specialty silicones. https://www.csb.gov/ab-specialty-silicones-llc/ (accessed 10 May 2023).

# 2

# Mechanisms of Accident Occurrence

## 2.1 Introduction

At the time of writing this book, on the night of 2 January 2024, two aircraft collided with each other on Runway C at the Tokyo International Airport or commonly called Haneda Airport, Tokyo in Japan, which was crowded with passengers returning home for the New Year and tourists. Both aircraft were totally damaged in a fire that broke out after the collision, leaving only the solid equipment, including the aircraft's frame, remaining. In this accident, a Japan Air Line (JAL) aircraft, immediately after landing on Runway C after receiving permission to land from Air Traffic Control (ATC), collided with a Japan Coast Guard (JCG) aircraft that had entered Runway C and was stopped on the same runway for approximately 40 seconds to take off. The ATC instructed the pilot of the JCG aircraft to proceed to the stop line on the taxiway and in the context of the term "Number one" being used together, meaning the order in which the aircraft was to take off. The pilot of the JCG aircraft assumed that he had received permission from ATC to enter the runway and proceeded with the aircraft onto Runway C. This accident was reminiscent of the Tenerife tragedy described in Chapter 1.

Five of the six JCG crew members were killed; the JAL aircraft carried 12 crew members and 367 passengers, including eight infants, all of whom escaped through two forward and one rear ejectable evacuation slides in an adverse environment where the aircraft was engulfed in flames, and the cabin was filled with smoke as a result of the collision. News broadcasts around the world hailed the escape of all passengers and crew from the JAL aircraft, describing it as a miraculous escape.

However, the miraculous escape of all passengers by the cabin attendants on the JAL aircraft involved concepts derived from accident models developed since the Tenerife tragedy. The concept is an approach of building multiple layers of defenses such that a single error does not lead directly to a catastrophic or serious accident. In line with this, accident prevention findings and lessons learned, based on scientific accident investigation analyses conducted by national accident investigation bodies, have been embodied in hardware and software aspects by aircraft manufacturers, airlines, national aviation authorities, International Civil Aviation Organization (ICAO), and other relevant organizations, such as aircraft equipment, aircraft materials, and granting passenger ejection authority to cabin attendants. The author, in the past, had participated in training conducted by JAL cabin crew to eject passengers using emergency ejection slides. The training was conducted using a training aircraft of the same dimensions as the actual aircraft, and the process of guiding passengers and ejecting them using the emergency ejection slide was conducted in a realistic manner. This is the software layer of defense within multiple layers of defense.

*Accident Prevention and Investigation: A Systematic Guide for Professionals, Educators, Researchers, and Students*, First Edition. Koji Fukuoka.

As part of the approach to preventing accidents occurring at universities etc., it is important to build multiple layers of defense to ensure that a single error by a researcher does not lead directly to death or injury, as described previously.

This chapter discusses useful concepts of hazard and accident, the law of risk reduction or hierarchy of control, the concept of defenses in depth to prevent accidents, the difference between a near miss and an accident and the boundary between the two, and the lessons to be learned from scientific accident investigation results and how they can be used, using real examples.

The importance of the effectiveness of wearing personal protective equipment (PPE) in the laboratory, as indicated by guidelines and regulations, is shown in relation to the concept of hazard. In addition, at the start of experiments in the laboratory or in the field, a wide range of safety measures and layers of defense are constructed to prevent people from being involved in accidents, such as the installation of alarms and the preparation of procedure manuals, and these layers of defense are discussed with real examples to illustrate their functions and importance.

In the safety field of other industries, when a near miss occurs that could have resulted in a serious accident, a thorough investigation and analysis of the near miss is carried out as an incident, and the resulting contributing factors to the occurrence of the near miss are used to prevent accidents. Thus, the investigation and analysis of near misses, which have the potential to cause serious accidents, is of great importance. The reasons for the importance of near misses and the difference between near misses and accidents, i.e. whether the mechanism of an accident ultimately stops at an incident with no casualties or whether it develops to the end and results in a serious accident, are described with real examples. In the absence of scientific and exhaustive accident investigation reports of near-miss incidents in universities and other academic institutions that could contribute to this near-miss topic, actual marine accidents are discussed in this section.

## 2.2 Concept of Hazard and Accident

Before carrying out potentially hazardous activities, such as experiments with chemicals that may cause fire or explosions, or field research activities at sea, researchers are instructed by the university authorities or the faculty member in charge to carry out a thorough risk assessment in accordance with the Occupational Health and Safety Association (OSHA) regulations and other relevant regulations. However, reports of investigations and analyses of accidents that occurred at universities indicate that general safety training was provided, but education and training to assess the risks specific to the experiments was not carried out (The US Chemical Safety Board [CSB] 2010).

The concept of hazard is useful when considering the process of how a hazard affects people, e.g. personal injury, health damage, and its prevention. According to ISO/IEC Guide 51 (1999), a hazard is a potential source of harm. A hazardous situation involves presence of a hazard in which people, property, or the environment coexist. The process from the coexistence of a hazard to an accident is as follows. In a situation where people, property, and the environment coexist with the hazard, a person performs a hazardous behavior, which is termed as a harmful event in the ISO/IEC Guide 51, such as an error, mistake, or rule violation, which results in damage, which is named as harm, to people, property, and the environment. Figure 2.1 shows that layers of defense to avoid accidents or mitigate damage can exist as events or situations progress to the next process.

This process can still be applied to the occurrence of many events. However, in many cases, the theory of the Swiss cheese model and the systemic accident model, which will be discussed in Chapter 3, are applied to accidents that occur in modern society where the function of each element or part of a society, organization, or system has become more complex.

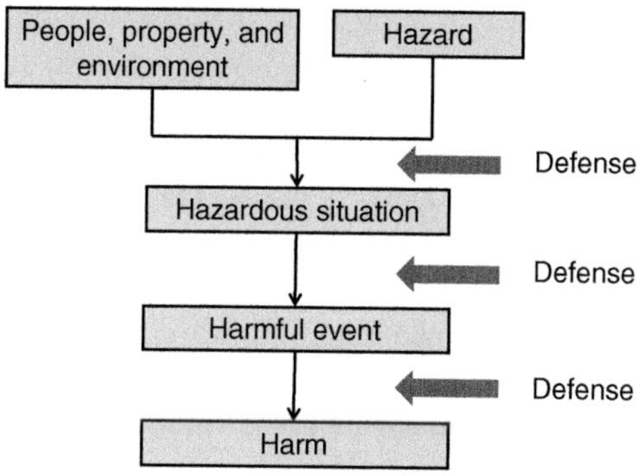


**Figure 2.1** Concept of hazard and accident (Fukuoka 2019).

### 2.2.1 Handling of Chemicals and Concepts of Hazard

Let us apply the concept of hazard illustrated in Figure 2.1 to accidents during the handling of chemicals. The researcher disposed of a chemical in a container with a water-reactive, flammable substance without considering the properties of the chemical, which ignited when it reacted with the water in the container, which was present due to the moist paper inside.

Water-reactive substances do not constitute a hazard on their own. For example, a water-reactive substance cannot be a hazard if it is stored in a chemical storeroom in a water-free bottle and in the absence of water-producing substances in the vicinity. However, it becomes a hazardous situation if a researcher or student takes the same chemical out of storage for an experiment and brings it into a laboratory where water-containing substances are present. The situation becomes one in which a hazardous, water-reactive substance and people coexist.

One approach to preventing accidents in this process is to isolate the water-reactive substance from water, i.e. to conduct the experiment in a glove box that does not contain water. Another approach is to use safer substances in the experiment, if feasible, instead of using the water-reactive substance itself, if the experiment is deemed to have the same effect. This approach may be possible in the chemical industry, for example, but if the research is carried out by universities or academic institutes whose aim is to create new types of value by exploring unknown areas, there are likely to be cases where the most effective method of risk reduction – intrinsically safe design cannot be applied, namely by substituting chemical substances. In such cases, the following risk reduction methods would be used.

Harmful events occur due to human error, mistakes, and other human activities, and the preventive measure is to create a system that does not allow errors, etc. to occur. When handling hazardous substances, researchers are initially aware of the risks and carefully prepare safety measures before conducting experiments. However, after handling the same substance a number of times, they may become less aware of the risks and scale up the experiment, assuming that because the experiment has gone well in the past without any accidents, it will be safe this time. It is necessary to ensure that a risk assessment is carried out, that the experimental procedure is documented and clearly stated as a protocol, including where in the process awareness should be focused for safety, and that the process is checked against this protocol during the experiment. If it is not practical for the researcher to verify the process against the protocol, they should avoid conducting the experiment alone and instead allow a third party to monitor the process.

The US Occupational Safety and Health Administration (OSHA) recommended in 2016 that experiments should be conducted by two or more people following an explosion that occurred during a solo experiment at a university. The National Academy of Sciences (2011) states that experiments whose work procedures are dangerous should not be conducted alone. It further states that if an accompanying researcher or "buddy" is not available, the experiment should not be conducted due to its high risk. The reason for this is that if an emergency situation arises, the researcher will not be able to respond adequately alone.

In addition to the reasons given above by the National Academy of Sciences, the following can be said about types of error. The findings of human error have scientifically proven that human mistakes cannot be detected by oneself, but only by a third party (see Section 1.10.1 in Chapter 1). It should be noted that safety is not guaranteed simply by carrying out experiments with more than one person. Once the protocol has been established, the participants in the experiment must be adequately trained according to the organization's systematic education and training program. Through this process, buddy is able to detect errors during the experiment, or unsafe behavior that deviates from the protocol, and correct it to appropriate behavior.

### 2.2.2 Accidents Involving Anoxia and the Concept of Hazard

Some accidents in university laboratories involve researchers and students who become victims of anoxia. This section discusses a fatal accident that occurred at a university (Tokyo District Court 1994) involving a case of anoxia. To contrast the university accident, the safety measures taken in the industry to prevent accidents involving anoxia and gas poisoning are described in relation to the concept of hazard. In the maritime industry, the International Maritime Organization (IMO) developed standard operating procedures (SOPs) to prevent accidents such as anoxia in response to the many accidents that had occurred in the maritime industry where crew and workers working in enclosed spaces, such as cargo tanks and cargo holds on ships, had been victims of anoxia and gas poisoning accidents. The IMO distributed the SOPs to relevant organizations, associations, and shipping companies around the world to ensure compliance with the SOPs.

#### 2.2.2.1 Anoxia Accident at a University

In August 1992, a refrigerator used to freeze and store important experimental samples in a low-temperature laboratory (set at −20 °C) broke down due to a lack of cooling water. When a researcher realized that the refrigerator had broken down, he asked a repair company to fix it, but it turned out that it would take time to restore the machine, so he decided to inject liquid nitrogen himself without waiting for repairs. The failure of the refrigeration unit was not reported to the faculty by the researcher. The freezer contained valuable ice samples collected in Greenland, which were important to the postgraduate student as he was writing his PhD thesis.

The researcher and postgraduate student placed a total of 80 L of liquid nitrogen in the preparation room, which was the front room with an area of 6 $m^2$, to control the temperature rise in the cryogenic chamber. The preparation room was closed without ventilation equipment. The postgraduate student was aware of the dangers of liquid nitrogen, but he had to comply with the request of the researcher, who had been in the job for a long time, was senior to him, and had a hierarchical relationship with him.

The laboratory to which the postgraduate student belonged had no rules regarding the supply of liquid nitrogen to the preparation room. However, the general handling of liquid nitrogen was specified, and when liquid nitrogen was received from the liquid nitrogen supply station, the windows were to be opened and the ventilation fan was to be switched on to ensure safety.

Two people, the researcher and the postgraduate student, died of hypoxemia. The immediate cause of the accident was that the oxygen concentration in the preparation room fell below 18%, the concentration required for healthy human survival, when 80 L of liquid nitrogen introduced into the narrow, enclosed preparation room evaporated and the researcher and postgraduate student did not measure the oxygen concentration in the preparation room.

#### 2.2.2.2 Response to Anoxia Accidents Occurring in Industry

The tank hatches on each of the cargo tanks of a chemical tanker are installed to act as a defense between people, property, the environment, and a hazardous situation. In other words, the tank hatch acts as a physical barrier to prevent people from coming into contact with chemicals that can cause anoxia or gas poisoning. Therefore, as long as a person does not open the tank hatch or enter the cargo tank underneath, the person is safe and no hazardous situation will develop (Figure 2.2).

A hazardous situation is created when a person enters a cargo tank. The International Safety Management Code (ISM Code; IMO 1993) requires all ocean-going vessels operating in international waters to have a Safety Management System Manual (SMSM) on board. The SMSM requires the atmosphere in the cargo tank to be measured at various heights above the inner floor plate (i.e. the floor of the cargo tank) using a suitably calibrated gas/oxygen concentration detector before personnel enter an enclosed space such as a cargo tank. The reason for measuring at different heights is that gas/oxygen concentrations will have different values at different heights from the cargo tank floor due to the difference in density with air.

As a defense between a hazardous situation and a harmful event, the crew or cargo handling personnel should measure the atmosphere in the cargo tank with a gas/oxygen concentration detector and confirm that it is safe before entering the cargo tank to commence work. Measurements should be recorded. Workers will wear appropriate PPE depending on the level of the readings. In other words, a mechanism is built into the regulations to prevent a hazardous situation from developing into a harmful event.

The SMSM further incorporates a defense between a harmful event and the harm, shown in Figure 2.1, that prevents the event from deteriorating into harm. The master must have emergency equipment around the tank hatch and a crew member on deck in case of an event such as a person collapsing due to anoxia in the cargo tank. The crew member is required to maintain close contact with colleagues working in the cargo tank by means of a transceiver or other means of communication and to perform the monitoring function. In the event of an emergency, such as a person

**Figure 2.2** Tank hatches on chemical tanker. *Note:* The photograph shows a tank hatch on a chemical tanker being opened by a crew member. The tank hatch is a cylindrical structure on the deck of the ship, under which it is connected to the cargo tank. When the tank hatch is closed, the harmful liquid in the cargo tank cannot come into contact with people. The tank hatch is normally closed while the ship is underway and during loading and unloading and opened when loading and unloading are complete and the cargo tanks are cleaned by crew members. The same ship was involved in the same kind of accidents in 2010 and 2012, which resulted in the loss of crew members. *Source:* Adapted from the Japan Transport Safety Board.

falling into the cargo tank, the crew should inform the master. Together with other crew members, they measure the atmosphere in the cargo tank and then don the more stringent PPE to rescue the victim. However, the rescue team is limited to crew members who are trained and competent in rescue. For the safety management system (SMS), see Chapter 4.

## 2.3 Effect of a Layer of Defense

Figure 2.3 shows the priority order for risk reduction, based on ISO/IEC Guide 51:1999. Risk reduction measures must be taken in the following order: (1) step 1: inherently safety design, (2) step 2: protective devices, (3) step 3: information for safety, and (4) step 4: administrative controls and PPE. Measures (1)–(3) are to be taken by product designers and manufacturers, while measure (4) is to be taken by users who have purchased the product. Designers and manufacturers are obliged to take actions (1) to (3) in the order in which they should be taken.

Skipping (1) and doing only (2) and (3) would be a violation of international safety standards. If an accident occurs as a result of machinery or equipment that does not comply with international safety standards, the designer or manufacturer of the machinery is subject to product liability (PL) law and may be held liable (Mukaidono 2006). In addition, the national accident investigation authorities, which conduct scientific accident investigation and analysis as described in Chapter 1, investigate all goods and services related to accidents, and in the case of accidents involving machinery and equipment, they use scientific methods to analyze the problems and make safety recommendations to the designers and manufacturers.

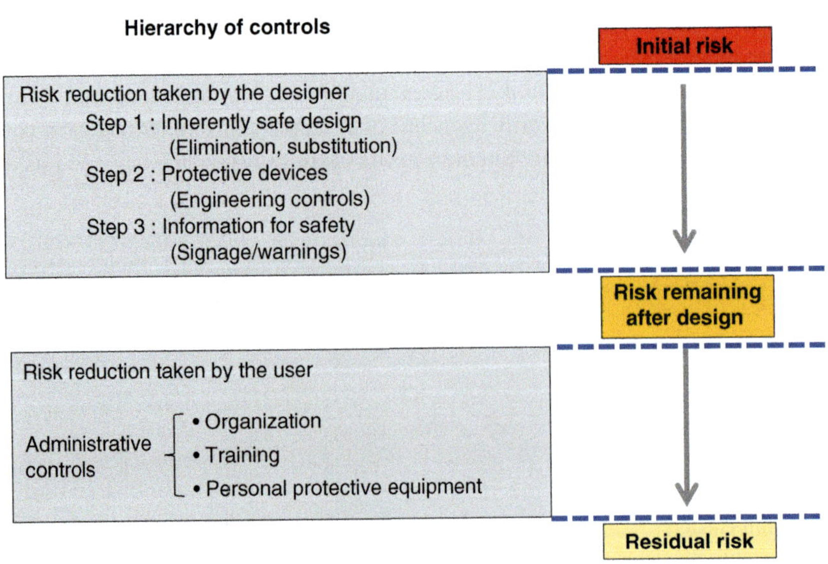


**Figure 2.3** Priority ranking in risk reduction methods. *Note:* This figure is a modification of ISO/IEC Guide 51 to help researchers in chemical experiments and field research activity to understand the effects of risk reduction. Designers and manufacturers carry out risk reduction by following steps 1–3 in sequence. This reduces the initial risk to the risk remaining after design. Users of equipment, devices, etc. are then required to write procedures, provide education and training, and wear PPE before experiments. However, even after PPE is worn, the residual risk remains when focusing on the risks shown on the right-hand side of the figure. *Source:* Adapted from Fukuoka (2016).

There are three types of measures taken by users of measure (4): organization, training, and PPE, but there is no order of priority between them. In this book, PPE is placed last for the following reasons, in order to be consistent with the experiments carried out in universities and other academic institutions.

- Experiments in university laboratories are carried out with PPE after organization and training before the experiment.
- Many accidents have occurred at universities where failure to wear appropriate PPE has resulted in personal injury, e.g. laser light damaging the retina of the eye, chemical substances penetrating gloves causing casualties, chemicals in the eyes, fire on clothing.
- It is important to show researchers and students that wearing PPE does not completely eliminate the risk and that there is a residual risk, although the risk can be gradually reduced by following steps 1–3 correctly.

The rules of the hierarchy of control can also be applied to experiments conducted in laboratories and field research activities at universities and other academic institutions. However, when applying the rules to experiments, the approach differs depending on the nature of the experiment and the substances handled in the laboratory or field research activity.

When conducting research on hazardous or unknown substances intended for chemical experiments, steps 1–4 must be followed by the laboratory. When conducting experiments using homemade research equipment, steps 1–4 must also be taken by the laboratory. However, when experiments are performed using commercially available equipment, such as laser equipment and machine tools, the risk reduction measures taken by the laboratory are limited and the rules of the hierarchy of control apply as they are, and the same step 4 is taken as that to be taken by normal users.

The International Organization of Standardization (ISO) and the International Electrotechnical Commission (IEC) have developed a system of international safety standards for machinery and equipment, with the ISO responsible for mechanical safety standards and the IEC for electrical safety standards. There are many other safety standards, including interlock standards, system safety standards and emergency stop standards for safety systems, and electrical equipment safety standards and general sensor safety standards for electrical systems, but ISO/IEC Guide 51 is the basic principle at the forefront of these standards.

### 2.3.1 Hierarchy of Control Methods

#### 2.3.1.1 Inherently Safe Design

Inherently safe design means designing for the absence of hazards from the design stage or designing for reduced risk where hazards are present. When a researcher is involved in the design and construction of an instrument, he or she will first consider inherently safe design when applying risk reduction methods. Inherently safety design methods include elimination and substitution, where elimination is the removal of hazards or potentially hazardous events at the design stage.

For example, in chemical experiments, if the research topic involves certain hazardous chemicals, such as chemicals that can explode in certain quantities, this applies to methods to reduce the risk of explosion by reducing the amount of the chemical being handled to a small amount. However, when field research is carried out at sea, accidents often occur due to strong ocean currents, tidal currents, and local currents influenced by the topography of the seabed. In the mountains, avalanche accidents have occurred repeatedly during winter snowfalls in streams with specific terrain and slope gradients of 30° to 40°. These facts allow researchers to study in advance where accidents have occurred in the past and to avoid these locations, trying to conduct the experiment in other places instead. In chemical experiments, where scientific accident investigation and analysis has shown that the accidental cause of the explosion was scale-up and was due to a significant increase in the experimental use of chemicals, the amount used can be changed to the minimum amount that will not harm people, thereby preventing the accident from recurring. However, in field research activities, environmental factors such as the characteristics of the mountain, the influence of currents and tides, transparency and weather conditions, such as fog and snowstorms, obviously have a significant impact on the occurrence of accidents, so that it must be borne in mind that when experiments are carried out in the same location, significant risks remain despite the measures taken.

Substitution is a method used when the material of an experiment can be replaced by another, different one that is relatively safe. In the case of chemical experiments, this means changing the type of chemical, for example, replacing the toxic chemical used by the researcher with a safer, less toxic alternative. When a chemical is used, it is replaced with a less hazardous substance, rather than using a substance that is self-igniting, water-reactive, potentially explosive in small quantities, or a flammable gas mixture, if the same research results can be achieved. Universities and other academic institutions are less likely to use this substitution approach as their aim is to research specific chemicals.

#### 2.3.1.2 Protective Devices or Engineering Controls

Next to inherently safe design, engineering control must be considered. In the field of machine safety, engineering control is referred to as safeguarding and can be divided into three types: guards, protective devices, and additional protective measures.

##### *2.3.1.2.1 Guards*

Guards are physical barriers on hazardous parts of machines and other equipment to isolate people from hazardous parts. The system is designed to keep people out of harm's way by preventing them from approaching the hazardous parts and keeping them away from the hazard. There are fixed guards, which require tools to open or remove, and movable guards, which do not require tools. They are fitted with doors, gates, or covers to the hazardous parts of the machine (Mukaidono 2006; Mukaidono 2007; Kadowaki et al. 2011).

##### *2.3.1.2.2 Protective Devices*

Protective devices are anything other than guards, such as interlocks and enabling devices. Isolation by physical barriers alone is not sufficient to ensure the safety of persons. In the case of machinery, there are times when a person may need to deactivate a protective device and approach dangerous parts themselves for maintenance or repair. Safety can be ensured by introducing a mechanism that stops the machine if the guard is opened. If the machine cannot be stopped suddenly by opening the guard, the mechanism should be such that the lock attached to the guard does not open until the stop is confirmed. When a person approaches the hazard, the energy of the hazard is dissipated (Mukaidono 2006; Mukaidono 2007; Kadowaki et al. 2011).

##### *2.3.1.2.3 Additional Protective Measures*

Even if a person attempts to use a machine in accordance with the operating instructions, errors may occur and the person may not use the machine as intended. In such cases, measures other than intrinsically safe design, guards, and protective devices are referred to as additional protective measures. These include emergency stop functions and measures to facilitate escape or rescue of trapped persons. These include securing escape routes and evacuation areas and means to reverse the rotation of the drive (Kadowaki et al. 2011). In the event of an emergency, it has the effect of stopping the machine or device and preventing people from being involved in an accident or, if people are already involved, reducing the damage to people. However, the effectiveness of the emergency stop button depends on the location of the button. A situation may arise where an accident occurs during a solo experiment and the researcher is involved in the accident but is unable to stop the machine because the button is located in a remote area. The mitigation of damage caused by the emergency stop button is conditional, e.g. if more than one person needs to carry out an experiment.

With regard to guards, in the case of chemical experiments, the dangerous parts used in the description of guards may be replaced by dangerous chemicals. If experiments are conducted in chemical hoods, the chemical hoods physically isolate people and potentially hazardous chemicals. Regarding protective devices, in the laser area, a system should be set up where the entrance to the laboratory where the laser beam is emitted has an automatic locking function to prevent people from entering from outside during the experiment. When a person enters the laboratory, the laser is deactivated, i.e. if the laser beam is a hazard, the hazard is deactivated when the person enters. The installation of protective screens or shields to prevent the laser beam from being directly in a person's eyes is also under engineering control.

Accidents have occurred on machine tools where people have been caught in rotating belts, gears, saw teeth, etc. during operation. To prevent people from being caught in rotating objects, measures can be taken to provide engineering control between the rotating object and the person. In other words, guards are installed to cover the entire rotating body, such as a belt, or to shield the surrounding area to ensure safety.

#### 2.3.1.2.4 *Information for Safety*

The next step after inherently safe design and engineering controls is information for safety. This step 3 means that the designer or manufacturer provides the user with the information necessary for the safe use of the machine or equipment in question by displaying markings, signs/pictograms, and written warnings. The purpose is to clearly indicate what risks remain in the machine/equipment even after step 1 and step 2 have been carried out, to alert the user and to reduce the risks through appropriate action by the user. This includes clearly indicating the training required to operate the machine and including an instruction manual with the machine sold. This is to ensure that the user uses the equipment correctly in accordance with the instructions in the manual.

#### 2.3.1.2.5 *Administrative Control*

After risk reduction in steps 1–3, administrative control is what users should do to ensure safety when using machinery, equipment, chemicals and so forth. Administrative control includes the following types:

- Organization means the formulation of procedures, the development of experimental plans and procedures for approving or changing experimental plans, and ensuring compliance with regulations and guidelines by conducting internal audits and workplace inspections. It includes the formulation of environmental health and safety policies and regulations, the formulation of general SOPs for the general handling of chemicals and other substances, and specific SOPs.

  Specific SOPs are those that are not specified in the general SOP or are more detailed arrangements of what is specified in the general SOP. In the case of chemical handling, each laboratory is expected to formulate and implement procedures that govern the handling of specific chemical substances that may cause fire or explosion during experiments. It also includes the formulation and implementation of an emergency plan that specifies the laboratory's response in the event of a fire, explosion, etc. For field research activities, this includes a diving plan, an experiment plan, and an emergency response plan.

- Training includes training based on a systematic program of education and training provided within a university or other organization in which the skills required by researchers and students to achieve the objectives of the research are identified and acquired, as listed in the SOP. Experiments involving scuba diving in the ocean require training appropriate to the depth of the water and the difficulty of the emergency response. For chemical experiments, practical training in risk assessment should be provided according to the risk level of the chemical.
- Before starting an experiment, these procedures are followed before donning PPE. PPE includes goggles and face shields to prevent chemicals and liquids from splashing into the eyes and face, masks to prevent the inhalation of toxic gases, gloves to protect the hands from heat and harmful substances, protective clothing to protect the whole body from chemicals and harmful substances, footwear and safety caps to protect the feet and head from falling objects, and hearing protection to protect hearing in noisy areas. When selecting PPE, it is important to ensure that the material and construction of the PPE fully take into account the characteristics of the chemical substance to be handled in the experiment. For example, impermeable gloves should be used when handling permeable chemicals. Chemical splash goggles should be selected when glassware is used under high temperature or pressure conditions, or where there is a risk of explosion (National Academy of Sciences 2011).
- Experiments in the laboratory or in field research activities can be initiated with residual but reduced risk through the processes of organization, training, and PPE.

#### *2.3.1.2.6 Role of PPE*

According to the CSB (2010) in university experiments, it is a common practice in chemical experiments to place a blast shield between chemicals and people as a method of minimizing damage in the event of fire or explosion during the progression from harmful event to harm depicted in Figure 2.1. Within PPE, goggles prevent eye damage when hazardous liquids are splashed in the face during experiments, protective clothing prevent burns to researchers when chemical reactions cause fires in the laboratory by catching clothing on fire, and gloves prevent skin injuries from accidental contact when handling liquids that can damage the researcher's skin.

PPE cannot be a defense between hazardous situations and harmful events. In other words, if inherently safe design was not taken in a hazardous situation, the risk itself would be too great and the risk reduction effect of wearing PPE would be small. The human body is directly exposed to the same level of risk that the same layers of defense are exposed to, and in the case of chemical experiments, the human body is directly exposed to the effects of toxic chemicals. It should be noted that PPE is a last resort in attempting to reduce the risk and that the risk remains even if PPE is worn.

### 2.3.2 Importance of Inherently Safety Design

To reduce accidents and incidents in universities etc., faculty members and students need to be aware of the concept of hierarchy of control. In other words, they should understand that they should try to reduce risks according to the priorities in steps 1–3. They also need to understand the importance of signs and warnings with text and symbols to warn of hazards, the importance of instruction manuals for equipment used in experiments, and that administrative control, including the wearing of PPE, is a less effective method of risk reduction.

An accident on a chemical tanker in 2010, in which a crew member died from lack of oxygen, is a real-life example to support the above theory (JTSB 2011). The only access to the cargo tank of a chemical tanker in oxygen-deficient conditions is a cylindrical tank hatch, which is a diameter of approximately 78 cm and a height of approximately 62 cm on the deck. It is not intrinsically safe and allows a person to open the hatch and enter even in oxygen-deficient or toxic gas–filled conditions. As a result, there have been many cases of people entering cargo tanks/holds in low oxygen or high gas concentration conditions and wearing inappropriate PPE when unloading cargo or cleaning tanks, resulting in the loss of life. In the past, coworkers attempting to rescue a person who had lost consciousness after entering a cargo hold was involved in an accident and suffered a secondary casualty because they had failed to measure the atmosphere and wear the appropriate PPE.

International efforts have been made by the IMO, the Marine Accident Investigators International Forum (MAIIF), and national accident investigation bodies to prevent the recurrence of such accidents, and the IMO (2011) adopted a procedure on enclosed space entry and published on its website the lessons learned from its member states accident investigations. It should be noted that MAIIF was granted Intergovernmental Organization (IGO) status by the IMO general assembly in December 2009. MAIIF, through the national accident investigation agencies, distributed leaflets on the dangers of enclosed spaces to all shipping companies and shipping associations worldwide to inform them of accident prevention methods. However, despite these measures, the same accidents continued to occur.

This means that the measures taken, such as the distribution of leaflets and public awareness campaigns, have not led to a dramatic reduction in the number of accidents because they are not in the category of inherently safe design and are classified as administrative control, which is the least effective. Administrative control is to ensure safety by requiring workers and others to comply with manuals developed by the organization or management and to wear PPE appropriate to the environment in which they work. This is a measure that relies on human memory, and if people forget to take action and errors occur, administrative control is ineffective. Therefore, taking into account human characteristics, an inherently safety design requires that a system be installed on board to open the tank hatches only when the atmosphere inside the tank is within an acceptable risk range (Fukuoka 2016).

## 2.4 Situation of Defenses in Depth and Accidents

### 2.4.1 Use of Accident Models

As technology has advanced and socio-technical systems have become more complex, the mechanisms of accident occurrence have also become more complex, as they can no longer be described by a simple domino theory. Scientific accident investigation analyses carried out after accidents have occurred have shown that the existence of an imperfect layer of defenses, where failures occur from the beginning of the system installation or during operation, and the surrounding conditions that triggered the human factors contributing to the occurrence of accidents, are deeply involved in the occurrence of many accidents and incidents. It is difficult to provide meaningful information on the relationship between the occurrence of accidents and their contributing factors because the concept of hazards and accident, as generally presented in Figure 2.1, does not clarify and visualize individual and specific in-depth defenses.

The Swiss cheese model developed by Reason (1997) in the 1990s has been used as a model applicable to organizational accidents in modern society and is now used as a background theory in aircraft and ship accident investigation. Organizational accidents are accidents that occur when a person belongs to a company or organization and is acting as part of the organization's business, and accident investigation analysis has clearly shown that the safety management system of the organization has an impact on the human element related to accidents.

According to Reason (1997), the model shows that in the accident mechanism, there are several layers of defense between the hazard and the damage or accident, known as in-depth defense. The role of the in-depth defense is to prevent accidents, but each of the layers of defense that make up the in-depth defense is not itself flawless and perfect. The Swiss cheese model explicitly states that although the in-depth defense should ideally be flawless and designers, manufacturers, operators, and safety managers, etc. cannot foresee all possible accident scenarios; therefore, deficiencies occur during the design and operational process phases of the product and services. This defect is known as a "hole" in defense, where the hole is not in place, but moves, changes size, opens, and closes. This hole is caused by active failures and latent conditions. Active failures are slips, lapses, mistakes, and violations by operators in hazardous conditions that lead to an "accident trajectory." Latent conditions are conditions that do not become apparent until an accident occurs and are latent, i.e. safety-related defects are already present when machines, equipment, SMS, etc. are designed and set up, but management and operators are not aware of these defects while operating machines, equipment, SMS, etc. Or management and operators are aware of the existence of the deficiencies but continue to use them without taking corrective action. Latent conditions include poor design, inadequate supervision, undetected defects in the manufacturing process, maintenance failures, malfunctioning procedures, and deficiencies in organizational structures to ensure safety. An accident occurs when the holes in each layer of defense are aligned and the accident trajectory passes through these holes.

The Swiss cheese model, like other accident models, has its own problems. These are discussed in more detail in Chapter 3.

## 2.5 Near Misses

### 2.5.1 Near-miss Theory

A near miss is an event that would have resulted in an accident by the narrowest of margins. Therefore, no human casualties have occurred, and it is not an accident. Incidents and near misses do not develop into accidents due to the function of in-depth defense, and their mechanism of occurrence is different from that of accidents. Figure 2.4 shows an accident, incident, and near-miss model (Fukuoka 2024) based on actual near misses that could have resulted in serious accidents.

An organization achieves its production goals through a combination of events and conditions. In some cases, accidents, incidents, and near misses occur, as shown by the arrows in Figure 2.4. An accident occurs when the SMS and risk management established by the organization do not work, and the trajectory of the accident penetrates all the multiple layers of defense. Incidents occur when the multiple layers of defense within an organization work to minimize the consequences of harm. Near misses, which could have resulted in serious accidents, are considered to occur when a third party, other than the organization to which the operator belongs, is involved and the layers of defense that the third party has learned and mastered are applied to the current event.

The circumstances under which a near miss that could have developed into a serious accident occurs are clear in the marine sector (Fukuoka 2019). The next section uses real-life examples to illustrate how the different mechanisms of occurrence of accidents and near misses that could have developed into serious accidents manifest themselves as events and contributing factors in the workplace. It also clarifies the boundary between a major accident and a near miss that successfully

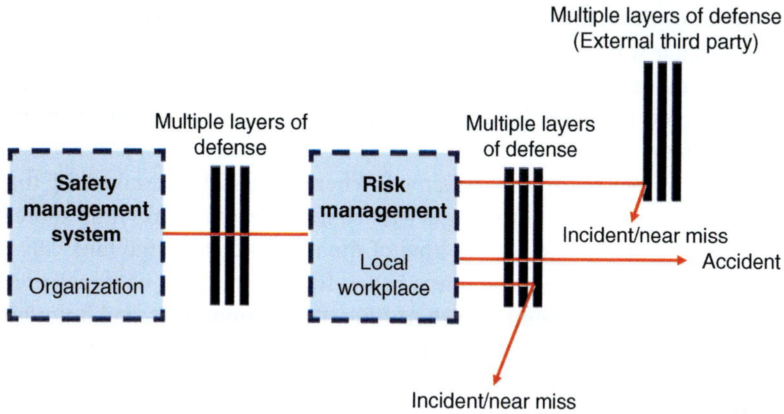


**Figure 2.4** Accident, incident, and near-miss model. *Note:* The red arrows show the trajectory of the accident, and the accident occurs by penetrating all the multiple layers of defense because SMS and risk management are not working. The dotted lines for SMS and risk management show that they are not working. It is assumed that a serious near miss is controlled by the multiple layers of third-party defense and the occurrence of a serious accident is avoided. *Source:* Adapted from Fukuoka (2024).

avoided leading to an accident, based on the near miss that occurred on the chemical tanker *Kyokuho Maru No. 2* (JTSB 2011) and an accident on the cargo ship *SINGAPORE GRACE* (JTSB 2012).

### 2.5.2 Hazardous Chemical Inhalation Accident: Chemical Tanker *Kyokuho Maru No. 2* Personal Fatality Accident

#### 2.5.2.1 Summary

The chemical tanker *Kyokuho Maru No. 2* was moored at the dedicated quay in Kawasaki Ward, Keihin Port, Japan, for unloading of the flammable liquid substance Tertiary Butyl Alcohol (TBA) at about 0155 on 10 March 2010, when the first officer, who was engaged in cargo handling operations, entered the cargo tank and was killed by oxygen deprivation.

#### 2.5.2.2 Events Leading up to the Accident

The chemical tanker *Kyokuho Maru No. 2* was in the process of discharging TBA from a cargo tank into an onshore storage tank using a cargo pipe when, halfway through the operation, the TBA could no longer be discharged. The first officer, on receiving this information from the chief engineer, looked into the cargo tank through a peep hole, or a small window, attached to the tank hatch, and found that a drain plug on the cargo pipe had been removed. The first officer donned a gas mask provided on board, opened the closed tank hatch by hand, and descended a vertical ladder into the cargo tank. The chief engineer, who was observing the first officer's activities near the tank hatch, noticed that the first officer, after reaching the bottom of the cargo tank and installing the drain plug, collapsed and became motionless approximately 10 seconds later. The chief engineer called loudly for the first officer, but he did not respond. The chief engineer saw the situation and reported it to the master. The master had three crew members prepare for the rescue.

#### 2.5.2.3 Circumstances of the Near Miss

A worker from the consignee, who was weighing liquids in an onshore tank near the jetty where the chemical tanker *Kyokuho Maru No. 2* was docked, rushed to the ship after hearing that the crew were in an uproar and that the first officer was unconscious and had collapsed. Around the tank hatch on the ship's deck, three crew members were seen trying to enter the cargo tank with a rescue line around their waists. The worker had learned about a similar case of this accident in training provided by the consignee and could predict that the cargo tank was oxygen deficient, so he stopped the three crew members, believing that they would die of oxygen deficiency if they put on gas masks and entered the tank.

As a result, secondary accidents were prevented. When the vessel was unloading the TBA into the onshore tanks by cargo piping, nitrogen gas was fed into the cargo tanks from shore to prevent explosion and cargo contamination. At the time of the accident, the cargo tank was filled with nitrogen gas and the oxygen concentration was below 10%. This oxygen concentration was such that a person could lose consciousness in a short time. It was found that, had it not been for the intervention of the worker, the three crew members would have been in the same situation as the first officer. The intervention of the worker allowed the three crew members to escape the oxygen deprivation and the event was a near miss for them, not an accident.

A decrease in the concentration of oxygen in the air is referred to as an oxygen deficiency, and if a person breathes air that is deficient in oxygen, he or she will suffer from an oxygen deficiency disease or anoxia.

#### 2.5.2.4 Properties of the Chemical TBA

According to the safety data sheet (SDS), TBA is a colorless, hazardous liquid that may cause respiratory irritation and requires the use of protective equipment such as gas masks when handled.

#### 2.5.2.5 Legislations

Article 50 of the Seafarers' Occupational Health and Safety Regulations states that if the shipowner allows the crew to work in areas where oxygen depletion may occur, the shipowner must measure the oxygen concentration before the work begins and ensure that the workers use breathing apparatus and other equipment. Under the legal system, the shipowner's obligations under the Seafarer's Act are assumed by the ship management company. The company that manages the ship and its crew is therefore obliged to comply with the above regulations.

### 2.5.3 Hazardous Chemical Inhalation Accident: Fatal Accident Involving Cargo Ship *SINGAPORE GRACE*

#### 2.5.3.1 Summary

The cargo ship *SINGAPORE GRACE* was docked at Saganoseki Port, Oita Prefecture, Japan for unloading of copper sulfide concentrate at about 0830 on 13 June 2009, when a worker collapsed while descending a ladder in the cargo hold to perform unloading operations. Two of the three workers, including the foreman, who knew of the worker's accident and went to rescue him, collapsed in the cargo hold, resulting in a total of three victims suffering from anoxia.

#### 2.5.3.2 The Events Leading up to the Accident

Before starting work, the foreman measured the oxygen concentration in the cargo holds with an instrument. The foreman recorded in the logbook that all the points measured had an oxygen concentration of 20.9%; however, the location of the oxygen concentration measurements differed from the method specified in the guidelines developed by the cargo handling company, and only some of the measurements were taken. The workers opened the hatch cover of the cargo hold to start work and used the crane to bring the heavy machinery into the cargo hold. At about 0830 on 13 June, worker A, who was climbing down the ladder from the lift entrance to operate the heavy machinery, collapsed in the cargo hold. When the foreman was informed of the accident, he, together with worker B and worker C, entered the cargo hold through the lift entrance to the cargo hold without measuring the atmosphere inside the cargo hold to rescue worker A (Figure 2.5).

Worker C followed the foreman and worker B as they descended the ladder from the lift entrance to the rescue without wearing the Self-contained Breathing Apparatus (a device that supplies high-pressure air under reduced pressure to the user), with the intention of stopping them. On the way down the ladder, the foreman inhaled oxygen-deprived air and collapsed in an anaerobic state. The foreman was unaware that the atmosphere was oxygen-deprived. Subsequently, worker C, together with worker B, descended the ladder from the lift entrance and reentered the cargo hold, wearing a gas mask provided by the crew, in order to rescue the foreman and worker A (Figure 2.6).

#### 2.5.3.3 Circumstances of the Near Miss

Worker C was advised by the cargo ship's crew that it was dangerous to enter the cargo hold, but as the mask provided was larger than a gas mask, he assumed it was an oxygen mask and thought it would be safe to wear it, so he and worker B went to rescue the foreman and worker A. Worker B descended the ladder from the entrance ahead of worker C, but on the way down he inhaled oxygen-deficient air and collapsed in an anaerobic state. Feeling weak and in danger on the way down, worker C struggled back to the entrance and reached the deck with the help of other crew members.

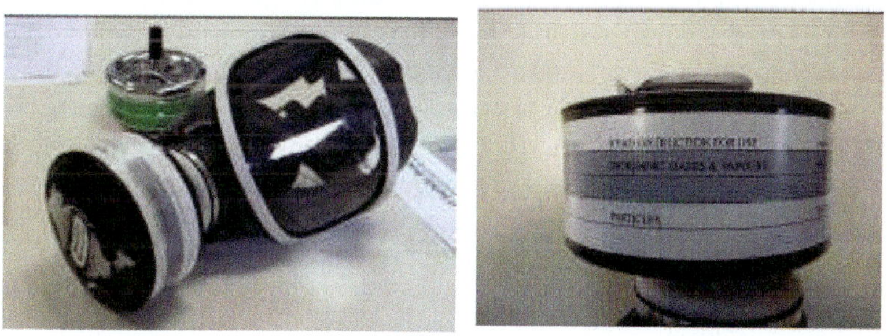

**Figure 2.5** Gas mask and chemical cartridge. *Note:* The gas mask and chemical cartridge are the devices worn by the victims; the chemical cartridge must be fitted with the appropriate cartridge depending on the target chemical, as the construction of the chemical cartridge differs depending on the type of gas. When using the equipment, the instruction manual states that the equipment must be used at a gas concentration of less than 1% and an oxygen concentration of 18% or more. *Source:* Adapted from Japan Transport Safety Board.

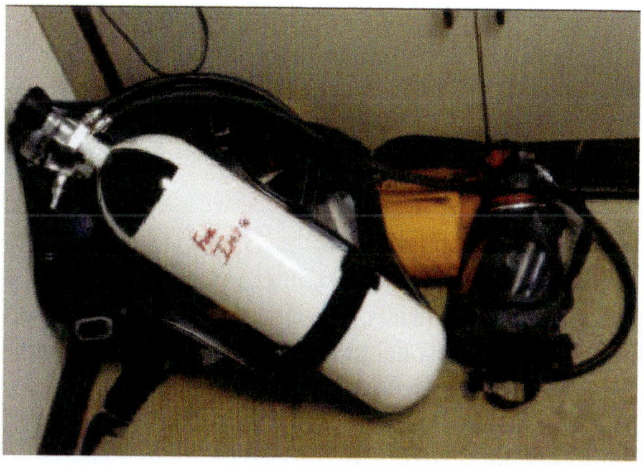

**Figure 2.6** Self-contained air breathing apparatus. *Note:* The breathing apparatus consists of a container and a facepiece that supplies compressed oxygen or air. *Source:* Adapted from Japan Transport Safety Board.

Worker C was aware that worker A had collapsed from lack of oxygen, and although he wanted to restrain the foreman and worker B, he felt he had to go to their rescue too, and driven by a sense of responsibility and frustration, lost his cool.

The cargo ship's side prepared a stretcher, an emergency escape breathing device, and a blower. The master was asked to allow the crew to go to the rescue with a self-contained air breathing apparatus but did not authorize the rescue by crew members as the cargo holds were not deemed safe. In addition to the crew members, the master, through the ship's agent, stopped the cargo handling company workers as they attempted to enter the cargo hold further.

#### 2.5.3.4 Causes of Oxygen Depletion

Immediately after the accident, the oxygen concentration in the lower part of the cargo hold, where the worker fell, was measured by the cargo handling company's workers to be between 1.2% and 2%. The oxygen concentration decreased the closer to the cargo surface. The following three factors were responsible for the extremely low oxygen concentration in the cargo holds at different locations:

1. Oxygen in the cargo holds was consumed as the copper concentrates oxidized and the atmosphere became oxygen-depleted.
2. The copper sulfide concentrates loaded by the vessel were chemically treated by flotation beneficiation. Flotation beneficiation is a method whereby copper ores with low copper content are chemically treated to produce copper concentrates. The method of flotation beneficiation is that the copper ore is pulverized and placed in water, to which oil and chemicals are added and stirred to collect the copper concentrates that adhere to the surface of the bubbles generated. Flotation beneficiator was used as the agent. In experiments using flotation beneficiator, oxygen concentrations were lower in the lower layers than in the upper layers.

   Therefore, when the copper concentrates with flotation beneficiate from steps 1 and 2 were stored under sealed conditions in the cargo holds, the oxygen concentration decreased in the lower part of the cargo holds, and the oxygen concentration in the cargo holds, which were oxygen-depleted due to oxidation of the copper concentrates, was different at each point where it was measured.
3. At the time of the accident, the workers opened the hatch covers to start work and natural ventilation was provided by the outside breezes. However, the wind speed of the outside air was low by 0–1.4 m $s^{-1}$ and convection currents were not strong enough to replace the air in the cargo holds with the outside air, so natural ventilation could not be achieved. The cargo handling company and the refinery did not see the need for forced ventilation as they had experience of oxygen levels rising to 20.9% with natural ventilation.

#### 2.5.3.5 Factors Contributing to Secondary and Tertiary Accidents

The foreman and two workers who died in the *SINGAPORE GRACE* cargo ship accident had been trained and had knowledge of the causes and symptoms of oxygen deficiency and its prevention, the use of self-contained breathing apparatus, evacuation and emergency resuscitation in the event of an accident, in accordance with the regulations for the prevention of oxygen deficiency and other oxygen-related diseases. However, they had not received adequate education and training on how to deal with personal injury in cargo holds where copper concentrates were being loaded. More specifically, they did not receive more practical education and training on what to do in the event of personal injury in cargo holds containing copper concentrates, such as measuring oxygen levels before starting a rescue operation, deciding whether or not to enter oxygen-deficient cargo holds, wearing protective equipment compatible with the atmosphere, carrying out systematic rescue procedures (Figure 2.7).

**Figure 2.7** Rescue situation by Fire Department Rescue Team. *Note:* The rescue team reduced the risk by using forced ventilation and oxygen cylinders to rescue the foreman and two workers who had collapsed in an anaerobic state before going to the rescue. The two people standing on the copper sulfide concentrate in the cargo hold are the rescue team. Next to the rescue team members is a ladder that the foreman and the workers used to descend from the cargo hold entrance. *Source:* Adapted from Japan Transport Safety Board.

### 2.5.4 Near Misses and Accident Boundaries in Serious Accidents

In the accident on the cargo ship *SINGAPORE GRACE*, the foreman and worker B, who attempted to rescue worker A, lost their lives in secondary and tertiary accidents. The foreman and all workers related to this accident knew the causes and symptoms of oxygen deficiency and its prevention, the use of breathing apparatus, evacuation, and first aid in the event of an accident, but they were not familiar with practical education and training. Worker C, who survived, recognized that he was suffering from lack of oxygen and lost his composure when he tried to stop the foreman and worker B from rescuing worker A. However, he felt weak on the way and turned back because of a sense of danger.

Meanwhile, in the accident on the chemical tanker *Kyokuho Maru No. 2*, the worker of the cargo handling company had learned about a case similar to the present accident as part of the education and training conducted by the company. The worker was able to stop the three crew members because he could foresee that the cargo tank was in a state of oxygen deficiency and that the crew would die of oxygen deficiency if they put on gas masks and entered the cargo tank.

In these accidents, the rescuers themselves were involved in the disaster when they attempted to rescue a person with limited knowledge of how to prevent oxygen deficiency and how to use air breathing apparatus in the event of personal injury. The conditions for effective risk control by an external third party and for correcting the trajectory of the accident to a near miss are considered to be met not only by the third party's knowledge of oxygen deficiency and its prevention but by learning the dangers of oxygen deficiency using real cases and how to deal with it from the lessons derived from the scientific accident investigation and analysis.

Although the discussion here is limited to the case of oxygen deficiency in other industries, it can be applied to the case of researchers and students who are fully aware and knowledgeable about the hazards of the chemicals they are handling but are still involved in disasters.

In other words, it is not enough to learn in education and training that a fire or explosion will occur when certain chemicals are mixed. Hands-on training in chemical laboratories based on actual accidents is necessary to learn which specific experimental processes increase the risk of explosion accidents and to learn specific methods to prevent such accidents. This allows the accident trajectory to be corrected to near misses.

The same applies to field research activities that involve diving in the ocean. Divers learn about the symptoms of decompression sickness, the mechanism of its onset and measures to prevent decompression sickness and other accidents from textbooks. However, even when these subjects are taught in the classroom, accidents still occur during diving. As with the chemical accidents described above, training based on lessons learned from actual accidents can be conducted in indoor pools, so that emergency situations can be avoided with near misses. These trainings could be made more effective if it were part of a systematic education and training program set up by an organization such as a university.

Rescue with a self-contained breathing apparatus is limited by the amount of air available and is considered to be possible for rescuing operations attempted in a short time. However, safety personnel at companies or universities should not carry out rescues unless they have undergone training to master the handling and fitting with a self-contained air breathing apparatus and how to approach to the accident site to carry out a rescue in a short time. Rescue operations should be left to specialist rescue services. It should be remembered that knowledge and practical training are two different things.

The approach that can turn an event that is considered to be a serious accident into a near miss that avoids human casualties is the result of the scientific and comprehensive investigation and analysis of many accidents in the past by our predecessors, as well as a characteristic of multiple layers of defense, where a single failure does not lead to a serious accident or human casualties.

## 2.6 Lessons Learned

In Section 2.4, a real case was presented in which the worker from the cargo handling company was able to prevent an accident involving three crew members because in an education and training program he had learned a lesson from similar accident case. Lessons learned refers to learning and training about when accidents occur and how to deal with them in order to prevent them, through the medium of scientifically and comprehensively investigated and analyzed accident cases. The reason why scientific and comprehensive investigation and analysis is a prerequisite for lessons learned is that partial investigations have proven to be ineffective in preventing recurrence, as demonstrated by the avalanche accident discussed in Section 1.11.

This section looks at explosion accidents that have occurred in universities. The type of accident is an explosion involving a solid (nickel perchlorate hydrazine: NPH) and a gas (flammable mixture of hydrogen and oxygen), demonstrating that although the chemical forms are different, there are many similarities in the factors contributing to the accident (Fukuoka and Furusho 2022).

Table 2.1 contrasts the accident contributing factors by reanalyzing the two accidents using the risk management and quality management process (RMQMP) approach model. This section describes the lessons learned from the scientific accident investigation analysis of the two explosions that occurred at the two universities.

For the RMQMP model, see the RMQMP model in Section 3.3.2 in Chapter 3, for background on its development, areas of application, and characteristics. For further details on its use in accidents at universities etc., see Chapter 5, Table 5.1 "Table of SHEL elements compatible with accidents in universities, etc. and associated safety recommendations." Table 5.1 is an improved version of Table 2.1 published in 2022. Improvements include the revision of each local workplace accident factor to show the causal relationship with the laboratory or organizational accident factor, and the addition of new safety recommendation topics that can be used as a reference when taking corrective action to prevent recurrence.

**Table 2.1** The accident-contributing factors by reanalyzing two explosion accidents using the RMQMP model (Fukuoka and Furusho 2022/Springer Nature/CC BY 4.0).

| SHEL element | Subdivision | Accident-contributing factors found at University A | Accident-contributing factors found at University B |
|---|---|---|---|
| Software | Procedure at the accident site | There was no protocol or standard operating procedure for synthesizing high-energy substances, and wearing personal protective equipment was left to the individual's discretion. | There was no protocol or standard operating procedure for working with explosive gas mixtures, and there was no usual practice of wearing personal protective equipment. |
| Hardware | Condition of equipment | NA | There was no grounding or bonding. A tank for dry air was being used. |
| Environment | Environment of workplace, such as a laboratory | NA | There was enough humidity to generate static electricity. |
| Central liveware | Psychological limits | Based on experience, it was found that a small amount of compound does not ignite or explode on impact even when it gets wet with water or hexane, and the risk of a large amount of NHP was thought to be similar. | Work dealing with high-risk substances and processes was not always recognized as high risk by many researchers. |
| | Knowledge, skill, experience, education, and training | Learning of techniques and methods associated with high-energy substances was completed with a literary review only; general laboratory safety training and education and training specific to this study were not undertaken. | There was a lack of awareness that a mixture of inflammable gases and oxidant gases could ignite due to an electrostatic discharge or metal friction when an incubator or bioreactor is opened. |
| | Communication | It was at the student's discretion regarding when to consult the principal investigator for experiment or scale-up changes. There was no related governing university policy.<br>There were two communication tools for group meetings and research notes, but they were mainly focused on experimental results and not used for safety purposes. | The researcher told the principal investigator that the thesis had discussions about the need for keeping the $O_2$ level within 4.0–6.9% and asked the principal investigator whether this factor should be considered and whether a fireproof lab coat was required, but (research institute) the principal investigator's response was not known. |

*(Continued)*

Table 2.1 (Continued)

| SHEL element | Subdivision | Accident-contributing factors found at University A | Accident-contributing factors found at University B |
|---|---|---|---|
| **Peripheral liveware** | Near-miss response | Two near misses occurred in the same research group three years before the accident. The first time, nitrogen was generated just before the completion of the reaction product and an explosion was heard. The second time, the students made a mistake when scaling up and set the measurement unit to 30 g. These near misses were not documented as lessons learned and were not disseminated to the entire university. | A day before the accident, electrostatic sparks were observed several times in ungrounded metal equipment, such as a cracked sound within the one-gallon pressure vessel when the same digital gauge's on/off button was pressed. However, the cause was not investigated and analyzed, thereby losing the opportunity to prevent the accident. |
| | Experiment discontinuation criteria | NA | If a high-risk situation is observed in the event of a near miss, all work, including that with high-risk chemical substances and processes, must be discontinued, and all procedures must be investigated thoroughly. |
| | Operational status of the safety management system | The EHS inspector had conducted a safety audit/inspection of 118 chemical laboratories prior to the accident, but the principal investigator did not take corrective action in many cases.<br>The chemical hygiene plan did not include comprehensive risk assessment guidance for the laboratories.<br>There was no obligation to undergo laboratory safety training that was provided online and in-person by the EHS staff. There was also video training offered for undergraduate students, but the need for hazard assessment before starting research in the laboratory was not mentioned.<br>There was no documented procedure and approval process for any changes to experimental plans. | The laboratory safety inspection conducted with the EHS's "Laboratory Safety Inspection Checklist" was not comprehensive and was reduced to a mere formality. The use of gas storage tanks was not confirmed, and it was necessary to have a section dedicated to compressed gas.<br>The chemical hygiene plan did not elaborate on how researchers should best handle safety regulations and practices.<br>There was no policy or procedure to ensure that laboratory-specific safety trainings for individual laboratories are conducted regularly.<br>Some laboratories lacked or did not have standard operating procedures for handling hazardous substances, implementing hazardous work or having protective barriers or emergency procedures. |

| | | | |
|---|---|---|---|
| | University environment | The inspection was reduced to a mere formality and carried out in the absence of the principal investigator because the principal investigator considered it a violation of academic freedom.<br>The EHS was stipulated to supervise chemical hygiene, but it was not under the authority of the vice president for research, and there was no authority to close the laboratory. | Lab safety inspection lacked cooperative aspects, such as inspecting the laboratory in the absence of the researcher.<br>Many researchers underestimated the chemical, biological, and physical hazards and the need for personal protective equipment. |
| | Involvement of external organizations affecting the safety of the university | The laboratory standards of the Occupational Safety and Health Administration did not address the physical hazards.<br>Funding agencies did not have a policy to limit the quantity of high-energy compounds that could be synthesized and lacked guidelines for assessing the risks in the laboratory. | Funding agencies did not demand risk analysis of studies that used explosive substances. |

*Note:* NA is an indication of accident-contributing factors that could not be analyzed because they were not included in the accident investigation reports.

### 2.6.1 Procedure at the Accident Site

At the time of the accident, neither university had protocols or SOPs for the specific chemicals and gas mixtures used in the experiments. In other words, at University A, the quantities of synthetic substances produced in the experiments were not documented or communicated; at University B, the procedure for producing the gas mixture at the lower explosive limit was not defined.

In addition, students at both universities were not wearing PPE appropriate to the risks they were facing in their experiments, despite the fact that they were handling explosive materials. More specifically, at University A, the wearing of PPE was left to the discretion of the individual and the postgraduate students were not wearing it at the time of the accident; at University B, there were regulations but PPE was not routinely donned in the laboratory.

#### 2.6.1.1 Lessons Learned from the Procedure at the Accident Site

- There is a need to establish rules and guidelines for the handling of chemicals that can be applied throughout the university. The rules and guidelines should include risk assessments, risk reduction measures, and lessons learned from case studies when handling chemicals.
- Chemicals stored and handled by the laboratory may not be able to address the risks associated with the handling of chemical substances with general SOPs or general guidelines and training based on these SOPs. Laboratory faculty and principal investigators (PIs) must establish specific SOPs or specific protocols for the substances currently handled by the laboratory and provide education and training to their research staff and students. This training should always be recorded with the name, date, time, and topic of the education and training, so that the faculty member or PI can verify the progress of the training. To this end, it is recommended that universities establish systematically structured education and training programs as a system. The level of skills acquired by researchers/students through the systematic education and training program should be equivalent to or higher than the risk level of handling chemicals in experiments, and the program should enable the faculty to verify this status.
- It is recommended that a safety officer is appointed in each laboratory to disseminate safety information to researchers and students and to confirm the obligation to wear PPE in the laboratory to reduce risk. The specific details of the PPE to be worn during experiments should be determined by the laboratory faculty and rules for wearing PPE should be established for each laboratory. Once the rules have been established, a space for wearing PPE can be provided at the entrance to the laboratory, and a procedure can be established to allow access to the compartmentalized laboratory after PPE has been worn.

### 2.6.2 Lack of Hazard Awareness (Psychological Limits)

At University A, several experiments with small quantities of nickel hydrazine perchlorate (NHP) had been performed prior to the accident and no unusual events such as explosions had occurred. The postgraduate students had no doubts about the scale-up and had not consulted the faculty or the PI in advance but had made their own decisions.

As for University B, when the accident investigation organization interviewed researchers and students in a number of laboratories about their risk perceptions, many of them did not perceive the substances or processes they were currently handling as high risk, although they were highly hazardous.

#### 2.6.2.1 Lessons Learned from Lack of Hazard Awareness

- Laboratory safety requires the development of experimental protocols and guidelines, as well as education and training, to ensure that the hazards specific to each research are correctly identified and that appropriate risk assessments are properly carried out.
- As OSHA (2013) recommends that experiments should not be conducted alone, a mechanism should be put in place for another researcher present to check the protocol against the actions and circumstances during the experiment to ensure that the procedure is followed and safety is ensured.

  In its analysis, the CSB notes that researchers generally start out meticulously and cautiously, but over time become familiar with the hazard and fail to perceive it correctly (CSB 2010). The technical term for this phenomenon is complacency (Parrott 2011), and one technique for preventing accidents caused by complacency is to always act on the facts rather than on assumptions.

  In other areas too, complacency leads to accidents, and mechanisms are built into operations to prevent them. In aircraft, for example, the pilot in command is called the Pilot Flying (PF). Next to him, the copilot acts as so-called Pilot Monitoring (PM), using instruments and visual observation to check the captain's piloting status and to monitor whether the aircraft is being maneuvered in accordance with the facts, guidelines, and regulations to ensure safety.

### 2.6.3 Communication

There was a lack of communication about safety between faculty or PI and postgraduate students or researchers. At University A, opportunities for communication between faculty member and postgraduate students included group meetings and research notes, but the focus was on the experimental design and results of the experiment and not on safety. At University B, the faculty may not have responded to questions posed by the researcher to the faculty about the safe handling of gas mixtures.

In a survey of university experiments conducted by the American Chemical Society (2016), of the 991 people surveyed, 70.5% of faculty and 52.1% of postgraduate students conducted experiments alone rather than in groups. According to Reason (1990), the mistake you make cannot be detected by yourself. Only a third party can detect your error. Therefore, if a researcher makes a mistake during an experiment carried out alone, the experiment may continue without the error being noticed, increasing the risk of an accident. Because of the findings of human error, experiments should be carried out by two or more people to prevent accidents.

#### 2.6.3.1 Lessons from Poor Communication

- It is recommended that researchers or students carry out experiments with more than one person. Based on the knowledge of human error, the risk of accidents can be reduced by conducting experiments with colleagues who are skilled at the subject of the experiment.
- From a laboratory safety point of view, it is beneficial to set an agenda for discussion of risk reduction of substances handled in experiments, e.g. at group meetings in the laboratory. For problems related to the handling of chemicals and other substances that cannot be solved within the laboratory, it should be possible to seek advice from researchers and experts doing the same research.

### 2.6.4 Response to Near Misses

At University A, three years before the explosion, there had been two near misses in the same research group, including excessive production of energetic substances. However, no investigation into the cause of the near misses was carried out in the laboratory and no information was shared among researchers. At University B, the day before the explosion, a researcher heard a cracking sound inside the pressure vessel when she pressed the same button as in the experiment at the time of the accident. In addition, static electricity was generated on several occasions when the hand touched unearthed equipment. Although these unusual events occurred, the laboratory did not investigate why these anomalies occurred.

#### 2.6.4.1 Lessons Learned from Response to Near Misses Including Criteria for Stopping Experiments

- When a near miss occurs in a hazardous situation, such as an experiment involving a chemical that can cause an explosion and fire, the experiment must be stopped and the factors that caused the near miss must be scientifically and comprehensively investigated and analyzed. What needs to be done after the investigation is the same as the corrective action taken after an accident: all factors that contributed to the near miss need to be corrected.

  Analysis of many accidents has shown that near misses occurred in unsafe conditions or hazardous situations before the accident occurred (Fukuoka 2019; Fukuoka and Furusho 2022). However, the near misses were not investigated and analyzed, and opportunities for accident prevention were missed (UC Center for Laboratory Safety 2016).
- Criteria for stopping experiments must be established to prevent serious accidents. The safety risk management and safety assurance processes introduced in Section 1.7.1 indicate that a risk assessment should be carried out again if the risk controls put in place are found to be ineffective or if new hazards arise. Resumption of the aborted experiment should take place when scientific investigation and analysis have identified the contributing factors to the near miss, and all these contributing factors have been corrected and a situation has been reached where safety can be assured.

### 2.6.5 Operation of SMSs

The SMSs at both universities were found to be weak. Workplace inspections did not examine the chemical-specific hazards stored in the laboratories and the level of risk posed by the chemical experimentation process. Deficiencies identified during workplace inspections were not followed up by the safety inspector or the Environmental, Health and Safety (EHS), resulting in a persistently unsafe condition where deficiencies were identified but no corrective actions were taken. The chemical hygiene plan, which corresponds to the SOP for chemical experiments, did not include a risk assessment for the laboratory and did not establish a policy and operational procedures for laboratory-specific safety training. The procedure for approving changes to the experimental plan was also missing.

#### 2.6.5.1 Lessons Learned from Operation of SMSs

- Chemical hygiene plans need to detail the risk assessment process so that it can be adapted to the individual risks in the laboratory. For this reason, the department in charge of formulating this plan needs to consult with faculty and experts to reinforce these contents.
- The chemical hygiene plan should include the functions of the SMS applicable to universities etc. Details are discussed in Chapter 4, but the following areas should be included: education

and training in the above risk assessment, preparation of education and training records, preparation of experimental plans and approval or modification procedures, purpose and role of workplace inspections and inspection methods, deadlines for correction of deficiencies identified during workplace inspections and follow-up actions, response to near misses and setting criteria for stopping experiments, investigation and analysis of accidents and near misses, and dissemination of lessons learned.

### 2.6.6 University Environment

The university environment is related to academic freedom and this aspect is unique to the university culture, which is not found in other industries. This culture manifested itself in behaviors, such as conducting workplace inspections in the absence of faculty or PIs, leading to mistrust and lack of cooperation between safety staff or inspectors and faculty or PIs.

#### 2.6.6.1 Lessons Learned from the University Environment

- It is recommended that the respective responsibilities and authorities of the safety inspector and the faculty member or PI are clarified and that safety and experimental information held by both are shared during workplace inspections. Safety inspectors should have the opportunity to discuss with the faculty member or PIs if they become aware of a safety issue in the laboratory so that the experiment can be conducted safely. The results and the safety of experiments should be communicated, together with real-life examples, in the induction course or education and training programs carried out after the recruitment of faculty members and researchers.

  In relation to the collaboration between safety inspectors and PIs, there is a survey on how researchers view safety regulations (Schröder et al. 2016). According to this survey report, 16% of university researchers said that safety regulations have a negative impact on productivity and 26% said that safety regulations hinder the process of scientific discovery. The results of this survey show that there is a significant gap between safety inspectors and researchers when it comes to safety and research activities. Despite the fact that one of the goals of researchers is to obtain experimental results while ensuring the safety of the laboratory, it was found that there was a lack of information sharing between the researchers and the safety inspectors and that there appeared to be an authority gradient between them.

  Scientific accident investigation analyses, such as those of aircraft accidents, have clearly shown that when there is the authority gradient within a team with the same objectives, useful information is not shared, leading to accidents. Therefore, to avoid this unfavorable situation, it is necessary to clearly document the responsibilities and authority of the safety inspector and the faculty or PI in workplace inspections, to share the safety and experimental information they both have, and to discuss all safety issues thoroughly so that the experiment can be conducted safely. Achieving experimental results and ensuring safety are not separate but integrated, and risk assessment must be an inseparable part of the experiment (Fukuoka and Furusho 2022).

### 2.6.7 Methods of Applying Lessons Learned

The following describes the methods for using the lessons learned from the accident analysis. Universities A and B shall rectify all identified contributing factors to the accident. By eliminating latent defects that were not apparent before the accident occurred, they can prevent the same accident from happening again.

Universities other than University A and University B can use the details mentioned in Table 2.1 to proactively prevent accidents in chemistry experiments at their own university. For example, other universities concerned about the lack of compliance with the wearing of PPE in the laboratory can refer to the lessons learned about the wearing of PPE listed in Section 2.6.1 and in subdivision on "Procedure at the accident site" in Table 2.1.

If EHS departments are unsure of how to respond to a report of a near miss in a chemical experiment from a laboratory handling hazardous chemicals, they can refer to the lessons learned on handling near misses in Section 2.6.4 to prevent a serious accident from occurring. Table 2.1 should be compared with the university's SMS, and if the SMS listed in the lessons learned is not in place, the university's safety-related system should be modified, taking into account the points in the Operation of the SMS section.

Importantly, many accident-contributing factors are common, even though the type of accident may differ. This means that other universities can reduce the risk of both chemical and mixed gas explosions by correcting these common accident-contributing factors (Fukuoka and Furusho 2022).

The use of the lessons learned described above has been one way of addressing individual issues. Universities and research institutions that wish to take an organization-wide view and manage safety as a system are advised to take the time to establish and operate the SMSs listed in Chapter 4.

## 2.7 Conclusion

This chapter uses the concept of hazard to look at the process from the situation where a hazard and a person come into close proximity to the development of an accident, and the existence of defenses that can prevent the development of each process. It then applied the hierarchy of control, used in industry as a method of risk reduction in risk assessment, to laboratory experiments and field research activities in universities and research institutes. In the hierarchy of control, it should be noted that there is an order of priority, that inherently safe design is the most effective in reducing risk, and that the wearing of PPE used in the laboratory is the last measure to reduce risk, but that the risk remains even after the PPE is worn. In addition, a real accident in which a person was killed by lack of oxygen was presented to demonstrate that machines and equipment without inherently safe design, which is step 1, will render the distribution of warning leaflets, the formulation of SOPs, and the wearing of PPE useless, no matter how many measures are taken from step 2 onwards.

It is noted that accidents that occur in modern societies with increasingly complex socio-technical systems have multiple contributing factors that cannot be resolved by concept of hazard or domino theory alone and that these contributing factors are usually invisible to the human eye and remain latent for a long periods of time. Furthermore, scientific and comprehensive accident investigation analyses by accident investigation agencies have revealed the following. Most of serious accidents, such as fires, explosions, fatalities, and injuries, are often caused by a single operator, induced to make an error by the surrounding artificially or naturally created environment. The error compromises the integrity of one layer of defense and then adds other layers of defense that were not previously of integrity. The error causes the layers of defense to collapse, as if a dam holding back water were to break, resulting in a serious accident. It was also shown that the only way to stop an error leading to a serious accident and to lead to a near miss with no human casualties is to learn lessons from the real accident and to train practically.

The reanalysis of the explosion accidents that occurred in the chemical laboratories of Universities A and B using the RMQMP model shows that, as in other industries, many accident-contributing factors lie dormant behind the explosion accidents. The difference between industrial accidents and university accidents is that industry has established a safety management system within the organization to ensure safety and has a web of layers of defense that have been proven to be effective as a result of past accident investigation and analysis. However, the SMS in universities is weak and lags behind the industry, as shown in Section 2.6.5 in this chapter. As a result, a single error by an individual in a university will likely continue to lead directly to a serious accident.

The introduction to this chapter discusses the aircraft collision that occurred at Haneda Airport in January 2024. In that accident, a fire broke out on the JAL aircraft, but all 379 passengers and crew escaped unharmed. Although the aircraft was a total loss of property, all lives were saved. As far as human life was concerned, the accident, which everyone who saw the footage of the crash and fire would have expected to be a catastrophe, was a near miss with not a single serious injury. This was the last stronghold of multiple layers of defense at work. The multiple layers of defense, built up through the efforts of the aviation industry, international organizations, government authorities, and accident investigation agencies, were indeed operating effectively, with a robust system in place to ensure that even a single error did not lead directly to a serious accident with enormous loss of life.

The way to prevent the recurrence of accidents in universities and other academic institutions, and to proactively prevent accidents from happening in the first place, is for top management, university administration, EHS departments, departments, and laboratories to work together with faculty members, staff, and students to build and operate multiple layers of defense throughout the university, so that a single error does not lead directly to a serious accident.

## References


American Chemical Society. (2016). *Guidelines for Chemical Laboratory Safety in Academic Institutions.* Washington, DC: American Chemical Society.

Fukuoka, K. (2016). Visualization of a hole and accident preventive measures based on the Swiss cheese model developed by risk management and process approach. *WMU Journal of Maritime Affairs* 15 (2): 127–142.

Fukuoka, K. (2019). *Safer Seas: Systematic Accident Prevention.* Boca Raton: CRC Press, Taylor & Francis Group.

Fukuoka, K. (2024). Forward-looking safety with the aim of developing scientific diving. Lectures presented at the symposium to educate about scientific diving in Fukuoka, Japan (6 June 2024).

Fukuoka, K. and Furusho, M. (2022). A new approach for explosion accident prevention in chemical research laboratories at universities. *Scientific Reports* 12 (1): 3185.

International Maritime Organization (1993). International Management Code for the Safe Operation of Ships and for Pollution Prevention (International Safety Management (ISM) Code). Resolution A, 741(18).

International Maritime Organization (2011). Revised Recommendations for Entering Enclosed Spaces Aboard Ships, Resolution A 1050(27).

ISO/IEC Guide 51:1999 (E) (1999). *Safety aspects: Guidelines for their inclusion in standards.* Geneva: International Organization of Standardization and International Electrotechnical Commission.

Japan Transport Safety Board (2011). Marine accident investigation report: Fatality of a crewmember onboard chemical tanker *KYOKUHOU-MARU No 2.* MA2011-7. (Japanese).

Japan Transport Safety Board (2011). Marine accident investigation report: Fatality of a crewmember onboard chemical tanker *KYOKUHOU-MARU No 2*. MA2012-4.

Japan Transport Safety Board (2012). Marine accident investigation report: Fatality of workers onboard cargo vessel *SINGAPORE GRACE*. MA2012-4.

Kadowaki, T., Fukuda, T., Hashimoto, S. et al. (2011). *Frontiers of Safety Engineering: Concept of System Safety*. Tokyo: Kyoritsu Shuppan Co Ltd. (Japanese).

Mukaidono, M. (2006). *Understanding how to Assess Risk*. Tokyo: Japan Industrial Safety & Health Association. (Japanese).

Mukaidono, M. (2007). *Basic Concept of Safe Design: ISO/IEC Guide 51(JIS Z 8051, ISO 12100(JIS B 9700)*. Tokyo: Japan Standard Association Group. (Japanese).

National Academy of Sciences. (2011). *Prudent Practices in the Laboratory: Handling and Management of Chemical Hazards*. Washington, DC: The National Academies Press.

Occupational Safety and Health Administration, USA (2013). Appendix A to §1910.1450-National Research Council Recommendations Concerning Chemical Hygiene in Laboratories (Non-Mandatory).

Parrott, D.S. (2011). *Bridge Resource Management for Small Ships: The Watchkeeper's Manual for Limited-Tonnage Vessels*. Blacklick: The McGraw-Hill Companies.

Reason, J. (1990). *Human Error*. New York: Cambridge University Press.

Reason, J. (1997). *Managing the Risks of Organizational Accidents*. Surrey: Ashgate Publishing Limited.

Schröder, I., Huang, D.Y.Q., Ellis, O. et al. (2016). Laboratory safety attitudes and practices: A comparison of academic, government, and industry researchers. *Journal of Chemical Health and Safety* 23 (1): 12–23.

Tokyo District Court (1994). Judicial decision of Tokyo District Court, Number (wa)16762 of the year 1994. (Japanese).

US Chemical Safety Board (2010). Texas Tech University Laboratory Explosion, No. 2010-05-I-TX. https://www.csb.gov/texas-tech-university-chemistry-lab-explosion/ (accessed 10 May 2023).

UC Center for Laboratory Safety (2016). Report to the University of Hawaii at Manoa on the Hydrogen/Oxygen Explosion of March 16, 2016: Report 2: Recommendations for Improvements in UH Laboratory Safety Programs.

# 3

# Accident Models

## 3.1 Introduction

This chapter discusses accident models that are important for understanding the mechanisms of accidents that occur in universities and other academic institutions and discusses the historical background to the development of these models, an overview of each model and the issues involved, and the models that can be applied to each industry, including universities, etc. (Fukuoka 2017, 2019). There are three broad categories of accident models: (1) sequential, (2) epidemiological, and (3) systemic. The article by Fukuoka and Furusho (2022) actually reanalyzed two explosions that occurred in chemistry laboratories at Universities A and B. The results show that there is a cause-and-effect link. The chain of events in the university explosion accidents was not linear. A comprehensive investigation of the factors contributing to the explosion accidents shows that there are numerous accident-contributing factors that are latent over a long period and are triggered by student errors in the laboratory, resulting in the collapse of several defective layers of defense at once and causing the accident to occur. For these reasons, it has been argued that the epidemiological accident model can be applied to university accidents. Note that the systemic accident model is excluded from the model because it argues that a cause-and-effect link is not necessary as a condition for an accident to occur.

In this chapter, the three accident models are outlined, and in particular, the epidemiological accident model applicable to university accidents is described in detail.

To give the reader an in-depth understanding of the application of accident models, six accident models are used to describe the accident at University A, which was included in the US Chemical Safety Board accident investigation report. They are domino theory, Human Factors Analysis and Classification System (HFACS), software, hardware, environment, liveware (SHEL)–Reason hybrid model, Australian Transport Safety Bureau (ATSB) model, and the Risk Management and Quality Management Process Approach (RMQMP) model. It was also reanalyzed using AcciMap, developed based on systems theory, and illustrated with diagrams to show the problems and differences that arise when each of these models presented in this chapter is applied to accidents that occur at universities.

The analysis method was based on the procedures and definitions used in each model after extracting all events and situations related to the accident in the SHEL model in order to homogenize the evidence collected during the investigation phase. See Chapter 1 for details of the accident at University A and the classification of accident-contributing factors and Chapters 5 and 6 for the development of safety recommendations.

*Accident Prevention and Investigation: A Systematic Guide for Professionals, Educators, Researchers, and Students*, First Edition. Koji Fukuoka.

## 3.2 Historical Background of Accident Models

Accident models are important as a means of understanding phenomena, such as accidents or potentially hazardous system behavior, and as a tool for sharing information about the factors behind accidents when communicating with others about accidents (Leveson 2011). According to Hollnagel (2004), accident models are a formulaic way of thinking about how accidents occur and include sequential accident models, epidemiological accident models, and systemic accident models. The principles, analysis objectives, characteristics, accident analysis process, and examples of typical models for each accident model are presented in Table 3.1 (Hollnagel 2004; Hollnagel and Speziali 2008; Fukuoka 2019).

### 3.2.1 Sequential Accident Models

The sequential accident model, also known as the domino model because it is represented by the domino theory proposed by Heinrich (1959), describes a linear progression from cause to effect, including the social environment, human defects, unsafe acts, accidents, and disasters. The mechanism of accident occurrence proposed by the model is that the domino blocks fall one after the other, starting with the social environment, until the disaster block collapses, eventually leading to the collapse of all the domino blocks, a situation that results in human death or injury. Preventing accidents is said to be possible by stopping one domino block in the middle of these from toppling over, in other words, by stopping one factor from occurring (Dekker 2006).

In the manufacturing of automobiles, confectionery, deep-frozen foods, and bottled drinking water, these products go through a series of processes and are carried out in an assembly-line operation by means of mechanization. When an event occurs in confectionery production, such as the presence of contaminants of machine oil in the product, a domino model can be used to trace the flow of work in sequence to identify the process in which the contaminants were mixed. This is a case where the domino model works well.

The domino model only assumes that a domino block is associated with one accident factor. If there is more than one accident factor in a domino block or if a domino block has multiple functions, the domino effect will stop there. This means that the domino model cannot be used.

Fault tree analysis (FTA), failure mode and effect analysis (FMEA), and why–why analysis are methods of analyzing accidents from the same perspective as domino theory.

In the 1970s and 1980s, there were a number of major accidents involving socio-technical systems, such as the Flixborough chemical plant accident in the United Kingdom (1974), the Three Mile Island nuclear power plant accident in the United States (1979), the crash of a Japan Airlines aircraft on Mount Osutaka in Japan (1985), the explosion of the Space Shuttle Challenger in the United States (1986), and the Herald of Free Enterprise accident in Belgium (1987).

After detailed investigation and analysis of accidents involving these organizations, it became clear that investigations and analysis could not be carried out using sequential accident models, including the domino theory, and that human error could only be explained by considering the circumstances surrounding the operator who made the error, for example, local workplace factors and organizational factors in the workplace. This necessitated a different type of accident model to the sequential accident model (Reason et al. 2006).

**Table 3.1** Classification of accident models.

| | **Sequential accident model** | **Epidemiological accident model** | **Systemic accident model** |
|---|---|---|---|
| Principle | Clear links between cause-and-effect. | Carriers, barriers, and latent conditions. | Close coupling and complex interactions. |
| Analysis objectives | Elimination and containment of causes | Strengthening defenses and barriers. | Monitoring and controlling performance variability. |
| Characteristics (advantages, disadvantages, applications, etc.) | The accident mechanism is suitable for the analysis of events where the occurrence of accident factors is linear, like the falling of dominoes, and where the cause and its effect are clear. The model can be used for engine failures in cars, failures in automated processes in the manufacturing industry, machine failures, etc. The model is limited in its scope of application because it cannot explain in detail the relationship between the organization and the occurrence of accidents and is not applicable to the analysis of accidents involving factors from complex socio-technical systems that are not linear, such as accidents involving ships, aircraft, and nuclear power plants. | The mechanism of accident occurrence is that active failures and latent conditions cause accidents to occur due to deficiencies in the many layers of defense that were already in place. The model captures not only the immediate cause but also the many contributing factors that led to the accident and provides a comprehensive picture of why the accident occurred. The model provides a concept of multiple layers of defense where human error does not directly lead to fatalities, fires and explosions, or serious accidents. It is used to analyze accidents in the marine, aircraft, automotive, and medical sectors. The model is not considered applicable to the accident analysis of certain complex socio-technical systems because of the need for a clear cause-and-effect link. | The mechanism of accident occurrence is not a cause-and-effect and epidemiological factor, but occurs in the interaction of the whole system. The model excludes cause-and-effect links and states that accidents occur in a complex balance of system components and those accidents are emergent properties rather than consequences. It states that accidents are a by-product of normal functioning and emphasizes the importance of continuous monitoring of the components of a socio-technical system to assess whether controls are working. It is suitable for investigating and analyzing failures of large communications facilities, missiles, rockets, satellites, and other failures. The model is not used by accident investigation agencies because it requires specialist knowledge, a great deal of effort and time, and because many accidents have cause-and-effect links. |
| Accident analysis process | Trace the development of events backwards, starting from where the accident occurred. | Trace the development of events backwards, starting from where the accident occurred. | Trace the development of events backwards, starting from where the accident occurred. |
| Examples of typical accident models | Domino model, fault tree analysis, event tree analysis, fishbone analysis, why–why analysis, failure mode and effect analysis, sequentially timed events plotting, etc. | Swiss cheese model, SHEL–Reason hybrid model, Human Factors Analysis and Classification System, ATSB model, etc. | Functional resonance accident model, systems-theoretic accident model processes, etc. |

### 3.2.2 Epidemiological Accident Models

Epidemiological accident models, such as the Swiss cheese model, have been developed as an alternative to the sequential accident model. The Swiss cheese model was developed in 1990 when James Reason was writing *Human Error* and was asked by the publisher to develop an accident model applicable to the many major accidents that were occurring at the time. The model was named the epidemiological accident model because the process of accident occurrence is analogous to the relationship between human health and pathogens.

The Swiss cheese model is now used to investigate, analyze, and prevent organizational accidents in shipping, aviation, railways, medicine, and other industries because it can explain the background theory and events that lead to organizational accidents in companies and organizations (Reason 1997; McNair 2007; Dell 2007; Davis et al. 2007; Landre et al. 2007; Salmon, Regan, and Johnston 2007).

The Swiss cheese model categorizes unsafe acts from a psychological point of view and introduces the concept of latent conditions. In other words, problems caused by human factors, organizational safety culture, etc. are latent conditions that undermine layers of defense or induce active failures (Reason 1990, 1997).

Epidemiological accident models include the Swiss cheese model proposed by Reason; the SHEL–Reason hybrid model (Japan Marine Accident Inquiry Agency 2008), which integrates the Swiss cheese model with the SHEL model; the socio-technical system model (Grech, Horberry, and Koester 2008), which was developed from the SHEL model; and the HFACS, which is a modification of the Swiss cheese model applicable to aircraft accidents (Wiegmann and Shappell 2003). In addition to these, there is the ATSB model (Australian Transport Safety Bureau 2008), developed by the ATSB and used to analyze ship, aircraft, and rail accidents.

The Swiss cheese model is used as a theoretical background for accident investigation and analysis by many accident investigation organizations worldwide, whether in the marine, aircraft, or rail sectors (International Civil Aviation Organization [ICAO] 1993; Marine Accident Investigators' International Forum 2015; International Maritime Organization [IMO] 2015).

Reason (1997) describes the occurrence of organizational accidents and holes in the Swiss cheese model as follows: an accident occurs when the accident trajectory penetrates holes when holes in the deep defense that exist between the hazard and the accident are aligned. Holes are caused by latent conditions or active failures. Latent conditions lie dormant for long periods of time without adverse effects and are caused by the inability of designers, managers, operators, etc. to predict all possible accident scenarios that may occur. Active failures are performed by the operator at the local workplace and can have a direct negative impact on safety in the system, which, unlike latent conditions, manifests quickly.

According to the theory of the Swiss cheese model (Reason 1990), accidents will not occur if a hole in any layer of defense is closed so that the holes in the deep layer of defense are not aligned and the trajectory of the accident does not pass through the hole. Holes are caused by latent conditions or active failures, and latent conditions facilitate the occurrence of active failures and undermine the mechanisms of the layers of defense.

Therefore, identifying the location of the hole and clarifying its relationship to latent conditions will help close the hole and thus prevent accidents.

### 3.2.3 Systemic Accident Models

Systemic accident models exclude the concept of accident causes, or cause-and-effect links, from the model and assume that accidents occur when the components of a socio-technical system change. They are represented by the functional resonance analysis method (FRAM;

Hollnagel 2013) and the systems theoretic accident model and processes (STAMP; Leveson 2011). The systemic accident model was developed because the epidemiological accident model cannot explain accidents that occur when normal people carry out the same tasks as usual by eliminating the cause-and-effect links where accidents occur and focusing on all the components that make up the socio-technical system approach. Accidents are then assumed to occur when problems arise in the interrelationships between the components.

Dekker (2006) points to the results of the accident investigation into the mid-air loss of the space shuttle Columbia in the United States on 1 February 2003 as evidence of the need for a systemic accident model. He states that although the physical causes of the mid-air failure can be explained by the sequential accident model and the epidemiological accident model, these models cannot explain the possible organizational underlying causes. In other words, it is claimed that the epidemiological accident model cannot explain why engineers have come to believe that behavior or system performance that deviates from the norm is normal or acceptable.

Dekker (2006) states that the systemic accident model has two characteristics. One is that safety is an emergent property, meaning that the components of the system fit well with other components and the whole works. The other is that although the interrelationships between the components are governed by controls, such as procedures and design requirements; the socio-technical system fluctuates, and the controls in place are not adapted to the latest situation.

According to Leveson (2011), the systemic accident model does not consider human error as the main focus of analysis, as human error is considered necessary for the development of the system and the achievement of its objectives. In addition, problems caused by human factors and the safety culture of the organization are analyzed within the framework of emergent characteristics and control, as shown in the STAMP analysis method in Section 3.3.3. The systemic accident model therefore differs from the sequential and epidemiological accident models in that it assumes that the occurrence of accidents is a by-product of normal functioning and emphasizes the importance of constantly monitoring components of the socio-technical system to assess whether controls are working.

## 3.3 Problems in Each Accident Model

### 3.3.1 Sequential Accident Model

The sequential accident model is characterized by the linear, sequential development of accident factors and can therefore only be applied to accidents or industrial sectors where the accident factors of the system can be clearly physically traced. The model is not applicable to accidents occurring in socio-technical systems where the interactions between the components of the system are complex and difficult or impossible to trace. If a single part or domino block has multiple functions, the sequential accident model cannot go beyond that point. In other words, it cannot be analyzed.

Leveson (2011) argues that although technological innovation has dramatically changed the safety-engineering environment, resulting in systems with complex relationships between components, engineers still use fault tree analysis and FMEA, which are best suited to analyze linear systems, to analyze accident factors, etc. She argues that for these reasons, safety engineering has become less effective and that probabilistic risk assessment has limitations in assessing risk in socio-technical systems because it uses linear sequential accident models to quantify hazards.

#### 3.3.1.1 Analysis of the Accident at University A Using the Domino Theory

Figure 3.1 shows a reanalysis of the explosion accident at University A based on the domino theory proposed by Heinrich (1959). The five blocks of dominoes are arranged from root cause to social environment, fault of person, unsafe act, accident, and injury in order of root cause. Applying the factors of the accident at University A to each block, the root cause was the lack of protocols and standard operating procedures (SOPs) in the laboratory to which the postgraduate student belonged, which was the social environment. Next, the postgraduate student prepared for the experiment on the assumption that the nickel hydrazine perchlorate (NHP) would not ignite or explode even if a large amount of NHP was used, based on his previous experience with small amounts of NHP. The postgraduate student then scaled up to use more than 100 mg and was injured when NHP exploded while stirring it with a pestle.

One method of preventing this explosion accident in domino theory is to remove or reinforce one of the domino blocks so that it does not topple over. A possible solution would be to exclude the postgraduate students who were injured as a result of the scale-up from participating in the experiment. Instead, other postgraduate students would carry out the experiment. Although the replacement postgraduate student may initially be hazard-aware and careful in handling the NHP, as pointed out in Chapter 1, the replacement postgraduate student may also become less hazard-aware over time due to complacency and repeat the same accident.

Other measures include the creation of protocols and SOPs that are root cause. However, even when these are in place, accidents still occur. The real fatality case of the cargo ship *SINGAPORE GRACE* analyzed in Section 2.5.3 of Chapter 2 should be reviewed. The cargo handling company to which the foreman who died from lack of oxygen belonged had prepared SOPs and manuals. The work on site was not carried out in accordance with the SOPs.

The domino theory does not allow other accident-contributing factors at University A to be found, including the operation of safety management systems (SMSs), near misses, and

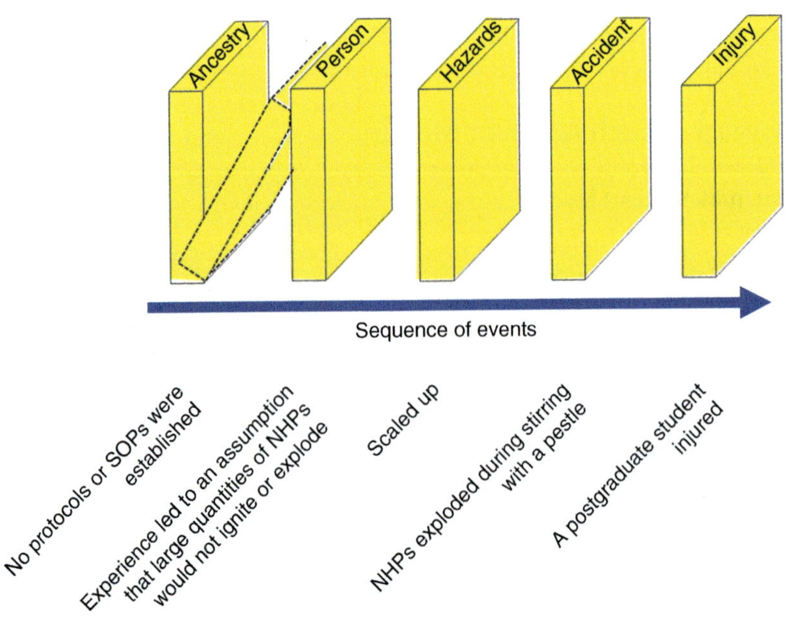


**Figure 3.1** Analysis of accidents at University A using the domino theory.

communication in the laboratory. Even if the domino theory is used to investigate, analyze, and take corrective action to address the problems, many accident-contributing factors will not become visible and will remain latent. Essentially, when accidents for which the sequential accident model cannot be applied are investigated and analyzed using the domino theory approach, only a subset of all the potential factors contributing to the accident can be visualized. The safety measures or safety recommendations derived from the investigation analysis would be neutralized, and the accident would recur.

### 3.3.2 Epidemiological Accident Model

In this chapter, the abbreviation SCM is used for the Swiss cheese model, and when referring to a model developed by Reason for a particular time period, Early, Middle, and Late are prefixed to SCM, such as Early SCM, Middle SCM, and Late SCM (Fukuoka 2017, 2019).

Issues with epidemiological accident models (Fukuoka 2016b) are discussed in relation to the characteristics of the HFACS, SHEL–Reason hybrid model, ATSB model, Swiss cheese model, Late SCM, and RMQMP models.

#### 3.3.2.1 HFACS

HFACS was developed by Wiegmann and Shappell in the aircraft sector to analyze military aircraft accidents and identify the factors behind the accidents. It has been applied to other sectors besides aircraft, such as shipping and railways, and is used to determine accident statistics and measures to prevent accident recurrence (Wiegmann and Shappell 2001, 2003; Schröder-Hinrichs et al. 2011). HFACS is based on the early Swiss cheese model published by Reason in 1990 (hereafter referred to as the Early SCM). HFACS has four layers of defense, the same as the SCM, standing upright, and shows that an accident occurs when the accident trajectory penetrates the four aligned holes formed in each layer of defense. The four layers of defense are classified into organizational influence, unsafe supervision, preconditions for unsafe acts, and unsafe acts of operators. These are further subdivided as follows:

- Organizational influence is divided into resource management, organizational climate, and organizational process. Resource management refers to the allocation of people, budgets, appropriate equipment, and facilities. Organizational climate refers to the structure, policies, and culture of the organization. Organizational process refers to the status of the operational plan, the presence of time pressures, and the development and monitoring of written procedures.
- Unsafe supervision includes inadequate supervision, planned inadequate operations, failure to correct problems, and supervisory violations.
- Preconditions for unsafe acts are divided into personal factors, conditions of operators, and environmental factors. Personal factors include teamwork, communication and operator preparedness, such as daily nutrition, education, and training. Conditions of operators refer to mental conditions, such as stress, complacency, and situational awareness; physiological conditions, such as fatigue; inability to process too much input information; and visual limitations. Environmental factors are the effects of the natural environment, such as weather, flight altitude, and route topography, and the interaction between the equipment and the operator, such as the human–machine interface.
- Unsafe acts of operators are errors and violations. Errors are classified into three categories: (1) skill-based errors, (2) errors that occur during decision-making, and (3) errors that occur when things are perceived. Violations are classified as routine violations and exceptional violations.

The problem with HFACS, as with the Swiss cheese model, is that the reasons why holes occur and the characteristics of holes, such as the movement of holes and the relationship between holes and accident factors, are not made explicit. Wiegmann and Shappell (2003) acknowledge the limitations of the Swiss cheese model for use in actual aircraft accident investigation and analysis and argue that holes need to be clarified if the model is to be used systematically. Reason et al. (2006) assessed that the Early SCM is not applicable to the analysis of complex organizational accidents because the development of organizational accidents is linear in the following order: wrong decisions, faulty line management, psychological signs of unsafe acts, unsafe acts, and inadequate safeguards.

##### 3.3.2.1.1 Analysis of the Accident at University A Using the HFACS

Figure 3.2 shows a reanalysis of the explosion at University A using HFACS. HFACS clearly defines and categorizes the breakdown of the four layers of defense so that it is clear which contributing factors fall under the heading of each layer of defense. The model was developed to fit the military aircraft sector based on the Early SCM and did not consider the influence of external organizations other than the military. Therefore, there was no subdivision corresponding to the involvement of external organizations affecting university safety, and there were accident factors that could not be extracted using the HFACS classification method. The main problem is that accidents are assumed to occur in the order of the four levels of defense.

#### 3.3.2.2 SHEL–Reason Hybrid Model

The SHEL–Reason hybrid model is an accident model that mainly combines the Early SCM and SHEL models with the Generic Error Modelling System (GEMS), which further classifies error types and violations, and was used by the IMO until 2014 as a guideline for marine casualty investigations.

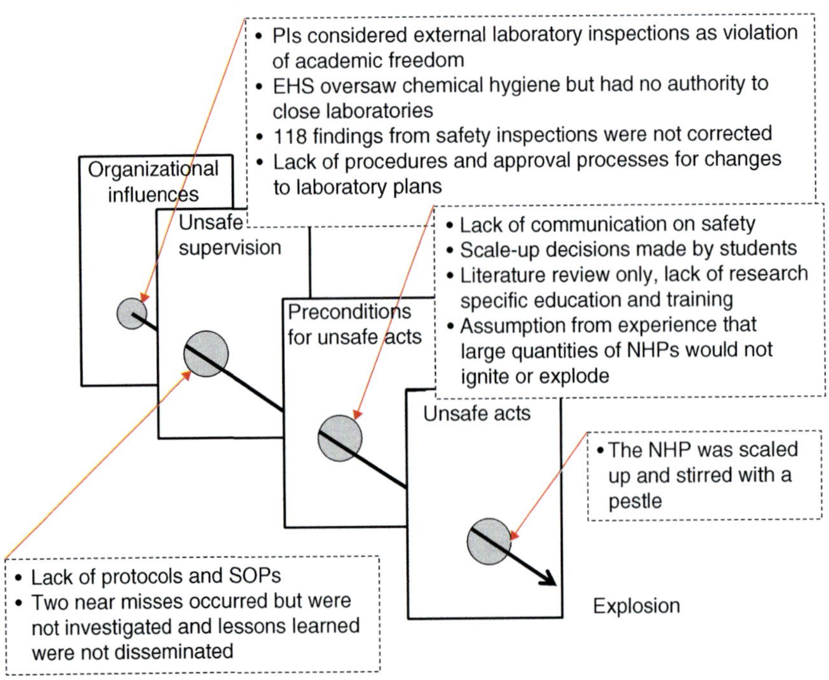


**Figure 3.2** Analysis of accidents at University A using the HFACS.

According to the same guidelines (IMO 2000), accident investigators are instructed to collect the evidence left at the accident site; to identify the circumstances leading up to the accident, the unsafe behavior of the operator, and the underlying factors influencing the unsafe behavior of the operator, i.e. latent conditions; and to take accident prevention measures to address safety issues.

The investigation process consists of the following six steps:

- Collection of evidence relevant to the accident

  A comprehensive collection of evidence left at the scene using the SHEL model should be carried out.
- Developing a timeline of events leading up to the accident

  The timeline is in the order of decision makers, line management, psychological signs of unsafe acts, operator unsafe acts, and inadequate layers of defense; each event or condition should be identified based on the information obtained from the data.
- Identifying unsafe acts and decisions and unsafe conditions

  Unsafe behaviors by operators are referred to as unsafe acts, which are errors or rule violations by operators in hazardous situations.
- Identifying types of human error or violation

  The types of unsafe acts and decisions are categorized according to GEMS as slips, lapses, mistakes, and rule violations. See Chapter 1 for more information.
- Identifying the underlying factors

  Identifying the factors behind individual or organizational behavior and decisions. Underlying factors exist in the local workplace and in the organization. The answer to the question "Why did you take this action or make this decision?" is the underlying factor. By comparing the elements of the SHEL model collected during the investigation phase, it becomes clear that there are underlying factors behind these behaviors and situations that lead to operator errors that occur at the local workplace. Examples of underlying factors are inadequate regulations and procedures, inadequate education and training, etc.
- Identifying potential safety problems and developing safety measures or safety recommendations

  Many safety issues are related to the organization's SMS. Safety recommendations are not developed for individuals because if the same individual changes to another person, it will be ineffective. This is because if another operator has received the same education and training as the operator who caused the accident, the same accident will occur even if the person is changed.

The problem with the SHEL–Reason hybrid model is that in analyzing accidents it assumes that the development of the accident starts with the decision at the top down to the operator and suffers from the same problems as the Sequential Accident Model, which appears to be linear.

#### 3.3.2.2.1 Analysis of Accidents at University A Using the SHEL–Reason Hybrid Model

The analysis of the explosion accident at University A using the SHEL–Reason hybrid model is shown in Figure 3.3, which provides direction for the planning of safety recommendations, as the unsafe acts are designed to specify the type of error and violation.

This means that in the case of violations, the focus can be narrowed down to the rules, while in the case of errors, the improvement can be targeted at the factors behind the inducement of the error. As the classification of each layer of defense is not defined in detail, there is concern that the contributing factors classified as layers of defense by different analysts may differ.

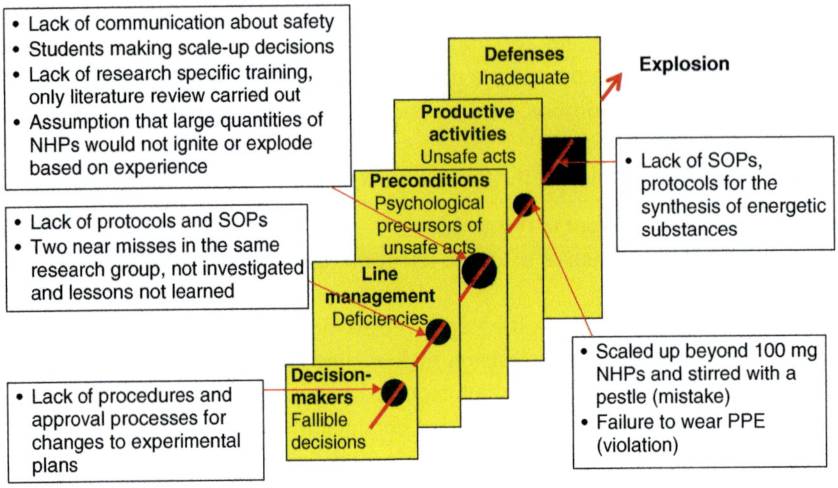


**Figure 3.3** Analysis of accidents at University A using the SHEL–Reason hybrid model.

#### 3.3.2.3 ATSB Model

The purpose of safety investigations by accident investigation agencies, including the ATSB, is not to question negligence but to determine the causes of accidents and to help prevent recurrence and mitigate damage. The ATSB model therefore eliminates terms used in the Swiss cheese model that are associated with negligence, such as failure and unsafe, and improves on the model presented by Reason (1995; hereafter referred to as the Mid SCM). Reason later modified these negligence-related terms.

According to the ATSB (2008), the ATSB model does not use terms such as cause, root cause, or proximal cause. The term "cause" is not used because the more complex the accident, the more difficult it is to determine cause, even after a thorough investigation, and because cause is a term that relates to responsibility and duty. Root cause and proximal cause are simplified and replaced by contributing safety factors. The contributing safety factor relates to the process leading to the occurrence of an accident rather than the end result of the accident.

The ATSB model looks only at the safety factors and identifies, from among the safety factors, the contributing safety factors and other safety factors involved in the accident. Among these safety factors, those related to the organization and the system are then considered to be of particular importance as safety issues and are the subject of safety recommendations. Safety issues do not include individual and local aspects.

Safety factors are events and circumstances that threaten safety, either in frequency or severity or both. A contributing safety factor is one which, if it had not existed, (1) the accident might not have occurred, (2) the undesirable consequences associated with the accident would not have occurred or would not have been significant, and (3) the other contributing safety factor involved in that accident would not have occurred.

The ATSB model consists of five levels, as shown in Figure 3.4, starting from the furthest from the accident site and working in order: organizational influence, risk control, local conditions, individual action, and the occurrence event. In order to derive the five levels of the ATSB model, it is necessary to use the appropriate analytical method to derive the factors for each level. In this regard, the ATSB introduces other analysis methods, such as barrier analysis, change analysis, behavior analysis, failure mode and effects analysis (FMEA) and gap analysis.

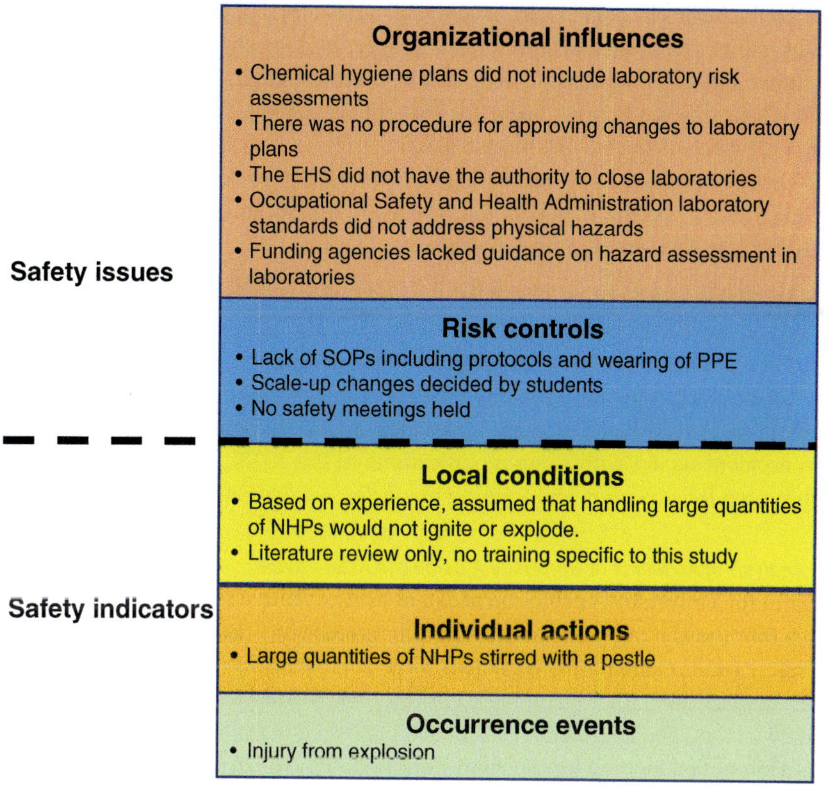


**Figure 3.4** Analysis of accidents at University A using the ATSB model.

- Organizational influences can be divided into two categories: risk controls implemented by the company and government regulations, international standards, etc. that affect the company.

  The risk controls implemented by the company include (a) resource management, including human resources, such as recruitment criteria, crew deployment and training, budgetary resources, such as cost savings, and equipment affecting the human–machine interface; (b) organizational culture, including chain of command structure, clarity of authority and responsibility, and management policies; and (c) organizational processes, including operational plans, the various procedures laid down in the SMS, internal and external audits, etc.

  Government regulations and international standards affecting companies include conventions and guidelines adopted by international organizations, such as ICAO and IMO, as well as safety regulations transposed from these conventions into national law.
- Risk controls are measures to prevent accidents or to reduce losses in the event of an accident and to minimize the frequency of deviations from the normal system. Examples include education and training, written procedures, bridge resource management (in aviation, cockpit resource management), communication and coordination (proper crew deployment), equipment backup systems, and alarm systems.
- Local conditions are those conditions that exist in the immediate vicinity of an individual's behavior or a technical event (breakdown). Specifically, they include: human factors, including the health, fatigue, stress, and alcohol consumption of operators, such as pilots, navigators, and vehicle drivers; factors such as the operator's experience and knowledge of the job; factors

caused by the job itself, such as excessive workload and time pressure; factors such as lighting, noise, humidity, vibration, and atmospheric quality; working environment, such as oxygen levels; weather and sea conditions, such as visibility, wind strength, wave height, snow cover, etc.

- Individual actions are actions taken by individuals (operators of aircraft, ships, etc.) that increase the safety factors or the risks to safety.
- Occurrence events show what has happened to aircraft, ships, trains, systems, etc. The behavior of individuals and the accidents that occur are inextricably linked. In other words, individuals increase risk, and accidents happen.

The ATSB model is based on the Mid SCM, which has been assessed as not being as applicable to the analysis of complex organizational accidents as the Mid SCM was to the Early SCM (Reason et al. 2006). On the other hand, however, Underwood and Waterson (2013) assessed that the ATSB model can be used as a systemic accident model because it incorporates a system thinking approach, a requirement that is classified as a systemic accident model, as described below. IMO (2014) has adopted an accident model with some modifications to the ATSB model in IMO Model Course 3.11, which is used for education and training in marine accident investigation.

#### 3.3.2.3.1 Analysis of Accidents at University A Using the ATSB Model

The analysis of the University A explosion accident by the ATSB model is shown in Figure 3.4. The advantage is that the actions taken to prevent a recurrence are clear, as the safety issues are the subject of safety recommendations. Weakness is the complexity of the work process, as new analytical tools are used to extract factors at each level and a relatively long process is required to identify safety factors. The modified ATSB model, published in the IMO model course mentioned above, uses a simplified method for the analysis tools and the process of identifying safety factors, which was previously cumbersome.

### 3.3.2.4 Late SCM

The Swiss cheese model introduced in Chapter 1, Section 1.8 is the latest in a series of models released by Reason, and the term Swiss cheese model now refers to the Swiss cheese model released by Reason in 1997 (hereafter referred to as Late SCM). According to Reason (1997), the Late SCM was developed by making two major changes to the Early and Mid SCMs. First, in the earlier model, the five layers of defense were distinguished by their names: fallible decisions by decision makers, deficiencies in line management, psychological precursors of unsafe acts, unsafe acts, and inadequate defenses. In the Late SCM, however, these names are no longer used and are collectively referred to as defenses in depth, which consist of multiple layers of defense.

The other is to replace the term latent failure or latent unsafe conditions with latent condition. Reason (2008) explains this change in terminology as unsafe is the result of a person's actions and is not an appropriate expression. The reason for Reason's change in terminology for unsafe is probably related to the fact that these terms are used after the outcome, such as an accident, has been confirmed and that operators would not knowingly perform an unsafe action while working.

Reason states that in Late SCM, holes are caused by active failures or latent conditions. Latent conditions are described as poor design, inadequate monitoring, undetected manufacturing process defects, maintenance failures, nonfunctioning procedures, etc. By not naming each layer of defense, Late SCM is considered applicable to accidents occurring in socio-technical systems, as it eliminates the linear process of accident occurrence, which was considered a limitation of the use of Early and Mid SCMs for complex accidents when they are applied to accident analysis and preventive measures.

As Late SCM is widely used internationally by ICAO, IMO, and others as a theory of accident mechanisms rather than a specific tool, no accident analysis of University A using the Swiss cheese model was performed here.

#### 3.3.2.4.1 *Challenges for the Swiss Cheese Model*

The Swiss cheese model, although modified between 1990 and 1997, has established itself as a scientifically useful model for analyzing and preventing organizational accidents by clarifying the existence of local workplace factors and organizational factors that underlie human behavior in relation to organizational accidents. The problem with the Swiss cheese model is that it is not clear what constitutes a hole, why a hole occurs, where the hole occurs, how it moves, how the hole opens and closes, and why the hole is aligned for an accident to occur (Wiegmann and Shappell 2003; Dekker 2006; Reason et al. 2006). These points apply equally to HFACS, the SHEL–Reason hybrid model, and the ATSB model, which were developed from the Swiss cheese model.

Regarding the problems with the Swiss cheese model, Reason (1997) states that the Swiss cheese model was developed to explain common concepts in organizational accidents. Dekker (2006) emphasizes that these issues are unresolved and states that it is left to the imagination of the accident investigator to decide how to view the problems with the Swiss cheese model. Reason et al. (2006) acknowledge that the Swiss cheese model is an established standard scientific method and is still used in many areas of accident analysis, that no significant experiments have been conducted that would negate the model, and that if latent conditions in the system can be identified in advance, accidents can be prevented from occurring in the first place. Therefore, in order to prevent accidents using the epidemiological accident model, it is necessary to be clear about where and when holes occur and what the relationship is between holes and latent conditions.

### 3.3.2.5 RMQMP Model

The RMQMP model, shown in Figure 3.5, was developed to explain the holes that have been a problem in the Swiss cheese model and the relationship between latent conditions and holes (Fukuoka 2016a,b). The model integrates the Swiss cheese model and SHEL models, risk management, and SMSs, and the diagram shows specifically where holes occur. In addition, a modified tabular format of the SHEL model makes it possible to identify the latent conditions that promote these holes during the investigation phase. The approach to accident prevention and proactive accident avoidance in this model is to construct multiple layers of defense that prevent a single error from leading directly to a major accident. It is believed that by correcting the higher level accident-contributing factors that make up each element of the SHEL model, shown in tabular form, multiple layers of defense can be constructed and accidents can be prevented.

Using this model, all accident-contributing factors can be visualized in tabular form, and accident-contributing factors can be statistically processed to generate accident occurrence patterns (Fukuoka 2016b; Fukuoka and Furusho 2016). Furthermore, it is believed that artificial intelligence processing of a large amount of accumulated data on accident-contributing factors will enable accident occurrence prediction and accident prevention measures.

The workplace uses the risk management process, and the organization uses the SMS (PDCA cycle) to ensure the safety of the whole system. Deficiencies that occur in these systems are indicated by red holes, which indicate at what stage of the process the hole occurred in the local workplace and organization. The accident-contributing factors or latent conditions that form the holes are listed in a tabular format. The tabular format has been modified to adapt each element of the SHEL model to university accidents.

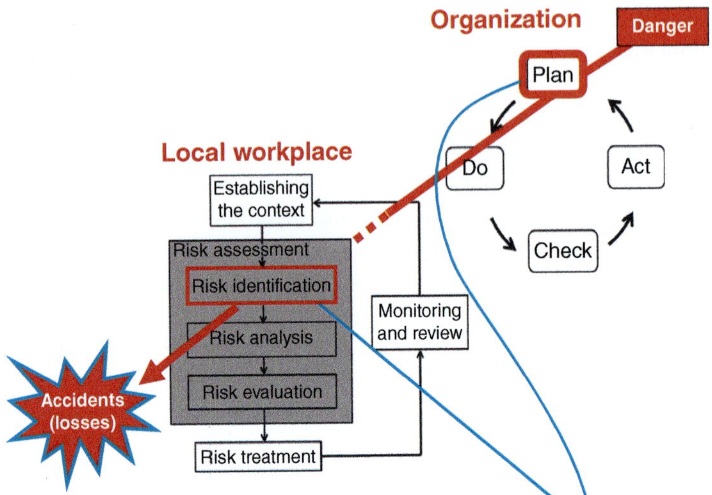


**Table showing contributing factors/latent conditions**

| SHEL element | Subdivisions |
|---|---|
| Software | Procedures, rules, and emergency responses at the accident site |
| Hardware | Conditions of equipment |
| | Human-machine interface |
| Environment | Environmental conditions of the workplace |
| | Meteorological and oceanographic conditions |
| | Traffic movement in the experimental site |
| | Geographical characteristics |
| Central liveware | Psychological/sensory limits |
| | Physiological situations |
| | Individual workload management |
| | Knowledge, skill, experience, education, and training |
| Peripheral liveware | Communication between experimental participants |
| | Team work |
| | Near miss response |
| | Experiment discontinuation criteria |
| | Operational status of the safety management system |
| | University environment |
| | Involvement of external organizations affecting the safety of the university |

**Figure 3.5** Analysis of accidents at University A using the RMQMP model. *Note:* The table showing contributing factors or latent conditions refers to Table 2.1 in Chapter 2. The locations of the holes in the layers of defense in the local workplace and organization and the accident-contributing factors that caused these holes, as determined by scientific and comprehensive accident investigations, are listed and visualized in the table.

The local workplace risk management process is an iterative process of hazard identification, risk analysis, risk evaluation, risk treatment, monitoring, and review to reduce the initial level of risk. During the accident investigation phase, it may be possible to determine the process in which the error or unsafe act committed by an operator occurred at the local workplace. Risk assessment at the local workplace corresponds to Endsley's (2000) situational awareness diagram discussed in Chapter 4. The Endsley Situational Awareness Diagram consists of level 1 (perception: awareness of current elements), level 2 (comprehension: understanding of the situation), and level 3 (projection: prediction of future situations), decision, action, and feedback.

The use of the model is closely linked to the actual accident investigation, with the following steps leading to the final safety recommendations (Fukuoka and Furusho 2022).

- After ensuring the safety of the local workplace where the accident occurred, evidence of the accident is collected using the contributing factors listed in Table 2.1 in Figure 3.5. As a preliminary step, an overview of what type of accident occurred, where, and how it occurred can be obtained in advance to ensure efficient and rapid evidence collection.
- Interview the persons involved in the accident about the circumstances at the time of the accident. For the operator who triggered the accident, the accident investigators shall reconstruct the accident at the site of the accident, obtaining from the operator a description of the actions leading up to the accident and of the situation at the site of the accident and in the surrounding area.
- The reconstruction shows how the accident occurred, the time line, and the unsafe actions or unsafe conditions of the operator. A timeline diagram is created by arranging the timeline of events in chronological order from left to right. If several organizations are involved in an accident, timelines can be added up or down, and the points where time coincides can be drawn as vertical dashed lines, for example, to illustrate the relationship between several timelines.
- Identify where the holes occur in the risk management layer of defense in Figure 3.5. During the reconstruction or when interviewing the operator, listen carefully to the operator's statement and ask for more information about the risk management process. If they were not aware of the hazard at all, then the hole occurred during the hazard identification process.

  A hole in the risk analysis process occurs when the hazard has been identified but the level of risk has not been analyzed using tools, etc. in accordance with a guideline or industry accepted method, due to assumptions, complacency, or other effects.

  A hole has been created in the risk evaluation process when a decision has not been made on whether the risk is acceptable or not based on the results of the risk analysis or whether an accident is likely to occur. If the results of the risk evaluation indicate that an accident is likely to occur but no accident avoidance measures are taken or the accident avoidance measures are inadequate, a hole has been created in the risk treatment process. If risk avoidance measures are taken in risk treatment but the effectiveness of the measures is not monitored, e.g. changes in surrounding conditions or the emergence of new risks are missed, then a hole has occurred in the monitoring and review process.
- Local workplace factors and organizational factors are identified from the evidence. The local workplace factors are the software, hardware, environment and central liveware listed as accident-contributing factors in Table 2.1 of Figure 3.5.

  The organizational factors are the accident-contributing factors listed in peripheral liveware (Lp) that identify safety issues in the organization. Particular emphasis is placed on those items included in the SMS.
- Identify where the holes in the organization's SMS defense layer have occurred. In other words, identify the process in the PDCA cycle where the deficiency in the SMS occurred.

  It is a hole in the Plan process if the chemical hygiene plan (CHP), experimental or diving plan, work plan, SOPs, or other safety rules and guidelines necessary to achieve the operator's objectives have not been prepared. Do process means to execute the plan and SOPs prepared in the plan process exactly as they are, including education and training. If plans, SOPs, etc. have been prepared, but no education and training has been provided on these plans and SOPs, or if experiments and field research activities are not carried out in the local workplace according to the plans, this is a hole occurrence in the Do process.

A hole in the Check process occurs in the following situations: when safety officers or auditors have not inspected or audited the laboratory during workplace inspections to determine whether the laboratory is carrying out experiments or field research activities in accordance with plans, SOPs, or other instructions; when an accident or near miss has occurred, but an accident investigation and analysis has not been carried out; and when lessons learned from accidents that occurred at other universities are not used to prevent accidents at their own universities, for example.

A hole in the Act process includes cases where workplace inspections, audits, accident investigations, etc. reveal safety issues in current plans, SOPs, etc., but these issues have not been corrected and updated.

- Recurrence prevention measures are designed to develop actions (safety recommendations) to correct the contributing factors identified in Table 2.1 of Figure 3.5. As there are organizational factors behind the unsafe acts of individuals, recurrence prevention measures will not be developed targeting specific individuals. Safety recommendations will be developed with a focus on correcting deficiencies in the Lp that are organizational factors. In this process, the SHEL model analysis identifies contributing factors that are not considered to be causally related to the accident. Since the contributing factors are identified as SMS deficiencies as a result of the investigation and analysis and may be other types of accident-contributing factors, safety recommendations are also made for the same contributing factors.

#### *3.3.2.5.1 Analysis of the Accident at University A Using the RMQMP Model*

One factor in the accident was that postgraduate students in the laboratory were unaware of the hazards associated with the scale-up of NHPs, and the explosion occurred when a quantity of more than 100 mg was mixed with a pestle (Table 3.2). Important accident-contributing factors in the organization's SMS were the lack of written procedures for laboratory-specific risks in the SOPs and CHP. Thus, it can be seen that the holes in the local workplace arose in the hazard identification process and in the organization in the Plan process. The latent conditions or contributing factors that gave rise to these holes are listed in tabular form (Fukuoka and Furusho 2022). One advantage of this model is that the location of the hole in the system and all contributing factors necessary to rectify the hole can be logically identified.

With regard to the tabular form of the latent conditions or contributing factors that caused the holes, refer to Table 2.1 in Chapter 2. Table 2.1 has been updated in this publication by adding topics that will help to understand the causal relationship of each contributing factor at the local workplace and to formulate safety recommendations to prevent recurrence. See Table 5.1 in Chapter 5 for its contents.

**Table 3.2** Summary of accidents and causes at University A (Fukuoka and Furusho 2022).

| Type of accident | Chemical explosion |
|---|---|
| Accident summary | During an experiment to synthesize and characterize energetic substances, a student was stirring nickel hydrazine perchlorate (NHP) with a pestle when the powder exploded. Graduate student injured. |
| Cause of the accident | The powder exploded while being stirred with a pestle because more than 100 mg of NHP was used (after the accident, the usage limit was set at 50 mg). |

*Note:* This table provides an accident type, overview of accidents, and causes from the accident investigation analysis into local workplace and organization.

Unlike the AcciMap described in Section 3.3.4, the causal relationship of each contributing factor is not shown in Figure 3.5. However, it is significant that many accident investigation bodies, such as the MAIB in the United Kingdom, contribute to accident recurrence prevention by issuing safety recommendations for all contributing factors identified as a result of accident investigations, which is the same approach as the ATSB model. From the perspective of accident recurrence prevention, safety recommendations for all accident-contributing factors that go beyond causality are considered effective, in line with the theory of the Swiss cheese model (Table 3.2).

#### 3.3.2.5.2 Solutions to Swiss Cheese Model Challenges

Fukuoka investigated and analyzed actual ship accidents and proposed solutions to the issues of the Swiss cheese model. Solutions are described here in simplified form; see the author's doctoral thesis and article (Fukuoka 2016a,b) for more details.

- What constitutes a hole

  The as low as reasonably practicable (ALARP) principle was used to describe what constitutes a hole. In other words, a hole was defined as occurring when the risk is in an unacceptable region, which indicates that the risk is not justified for any reason. The holes were then specifically defined and presented for each type of accident.

  Fukuoka (2016b) defines holes in relation to ship accidents. Selected chemical-related ship accidents that occurred on board ships include the following. The definition of a hole in an accident is the entry of a person into an enclosed area, such as a cargo tank, without measuring the atmosphere (e.g. oxygen or gas concentration) in the cargo tank during loading or unloading. See the oxygen deficiency accidents that occurred on the chemical tanker *Kyokuho Maru No. 2* and the cargo ship *SINGAPORE GRACE*, as discussed in Chapter 2.
- Reasons why holes occur

  The reasons why holes occur in the Swiss cheese model are shown by identifying the relationship between holes and latent conditions. The latent conditions that give rise to holes were identified as the 10 accident-contributing factors obtained from an analysis of 89 actual marine accidents using the SHEL model. These include voyage plans, procedures, regulations, human–machine interface, equipment conditions, environment, operator conditions, communication, teamwork, and management. In accordance with Pareto's law, the main latent conditions were defined as those for which the actual number of each latent condition obtained from the accident investigation analysis by accident type was within approximately 80% of the total cumulative ratio. Based on the proportion of these 10 accident-contributing factors as factors to the accident, the patterns of circumstances in which holes occur for each type of accident were then clarified. In other words, real examples show that holes occur due to the pattern of circumstances in which the accident-contributing factors indicated for each type of accident occur. Types of accidents include collisions, grounding, and accidents involving casualties.

  As an example, the following real-life examples presented in Chapter 2 are shown. The oxygen deficiency casualty accidents that occurred on the chemical tanker *Kyokuho Maru No. 2* and the cargo ship *SINGAPORE GRACE* are considered to have been caused by a combination of the following latent conditions, which led to casualties or caused the holes to open.

  Failure of the operator to recognize the hazard in the cargo tank during loading and unloading by not measuring the atmosphere with oxygen detectors or other devices in conditions where the atmosphere, including the oxygen concentration in the cargo tank, was deteriorating. Lack of clarity or failure of crew members, operators, and others to fulfill their roles. Operators lacking experience, skills, knowledge, education, and training in loading and unloading safety. That the organization does not have a written procedure for safe loading, unloading, or confined space entry operations, or that where a written procedure exists, operators do not follow it.

- Where the hole arises and where it moves

  The risk management protection layer of the local workplace and the SMS protection layer of the organization were considered as the places where the hole occurs and where it moves. Reason argues that holes are caused by latent conditions when the system is launched, since humans cannot predict all possible accident scenarios that might occur. Alternatively, he argues that holes occur and move and that they develop unnoticed, or at least uncorrected, while the system is in operation. The results of the analysis of marine accidents confirm Reason's assertion that the location of holes varies and is not necessarily the same depending on the type of accident – collision, grounding, casualty, etc. and that holes are found moving through the SMS protection layer and the risk management protection layer processes.
- Holes must be in a line for an accident to occur.

  In the 89 marine accidents studied, it was found that the trajectory of the accident was deflected because two or more holes, through which the trajectory of the accident penetrated, could occur in a single layer of protection, and the holes could not be aligned in a single row. The results of this study differ from the theory of the Swiss cheese model, which states that an accident occurs when a hole in each protective layer is in a row and the trajectory of the accident passes through it. The implication of this result is that all holes in the protective layers must be closed to prevent accidents.

### 3.3.3 Systemic Accident Model

STAMP, the leading systemic accident model, is an accident model developed from systems safety for aerospace systems, such as military aircraft and ballistic missiles, with the addition of systems thinking and human factors engineering. According to Leveson (2011), an overview of the analysis methodology is as follows. Note that STAMP is suitable for analyzing failures and problems in the military and aerospace industries, as well as in financial and telecommunications organizations, and the accident at University A was not analyzed using STAMP, as it did not fall into any of these categories.

- Identify the systems and hazards involved in the loss.
- Clarify the purpose and safety control of the system components associated with the identified hazards.
- Identify the immediately preceding events that led to the loss and analyze the losses incurred at the physical system level. Identify contributing factors, i.e. ineffective operational control, physical failures, dysfunctional interactions, communication and coordination failures, unaddressed failures, etc.
- Identify inadequate management that contributed to the losses in relation to the high-level SMS that oversees operational and other management. Consider the content of the responsibility, the status of fulfilling that responsibility, the context and impact of the decision, and any flaws in the decision-maker's mental model.
- Examine the loss-related coordination and communication situation for all components of the system and identify the content of the system and safety management structures that have changed over time.
- For all components of the system, identify the safety requirements and safety constraints, the circumstances under which decisions were made, inadequate control actions, and mental

model deficiencies, and describe these in a diagram consisting of each component. The management system channels and communication channels between the components are then depicted with arrows.
- Take recurrence prevention measures to correct the deficiencies and inconsistencies in the system identified in these analyses.

Underwood and Waterson (2013) state that systemic accident models are used exclusively by researchers because they require a detailed investigation of the normal conditions of the socio-technical system, in addition to the conditions under which the accident occurred. They go on to assess that the use of the models is not widespread among national accident investigation agencies and corporate safety personnel, who have time constraints when preparing accident investigation reports.

### 3.3.4 AcciMap

AcciMap is a model developed by Rasmussen (1997) based on systems theory. The model consists of four layers that cover all the social elements involved in an accident and allow causal relationships to be identified. AcciMap has the advantage that the causal diagram of an accident can be viewed systematically, making it easier to understand the underlying factors that led to the outcome and their interrelationships.

When planning safety measures, if measures are taken against higher level accident factors based on causal relationships, this can be expected to have an impact on accident prevention not only in a company or organization that has caused an accident but also throughout the organization using the same SMS. However, as the model attempts to be applicable to a wide range of fields, the elements of each layer use general terms, making it difficult to extract the elements that apply specifically to each layer. In addition, there is a concern that, even for the same accident, there may be considerable variability in the content of the resulting AcciMap, depending on the accident investigation and analysis skills of the users of the model.

The layers are as follows. In this section, the lowest layer will be referred to as the first layer and the uppermost layer as the fourth layer for the purpose of describing each layer. The accident factors corresponding to each layer shall be derived by other analytical methods.

- The first layer describes the outcomes of the accident. If a fire occurred during the experiment, followed by an explosion, both the fire and the explosion should be recorded. Below that, the damage caused to people and objects should be entered.
- The second layer is to fill in the operator's actions that are causally related to the occurrence of the accident. The behavior includes errors and violations. In addition, the psychological, physiological, and physical factors that triggered the behavior are entered.
- The third layer fills in the accident factors relevant to the company to which the operator belongs. It includes management related to resources, such as budget, allocation of personnel, upgrading of equipment, risk management undertaken by the organization, development of SOPs, training, monitoring, such as workplace inspections and audits, accident investigations, and sharing of lessons learned.
- The fourth layer describes organizations relevant to the accident other than the organization to which the operator belongs. It includes inspection bodies, certification bodies, local administrative bodies, government regulators, international organizations, and private bodies that produce guidelines and regulations in specific areas.
- Factors listed in each layer that have a causal relationship to outcome are linked by arrows from the top to the bottom layer.

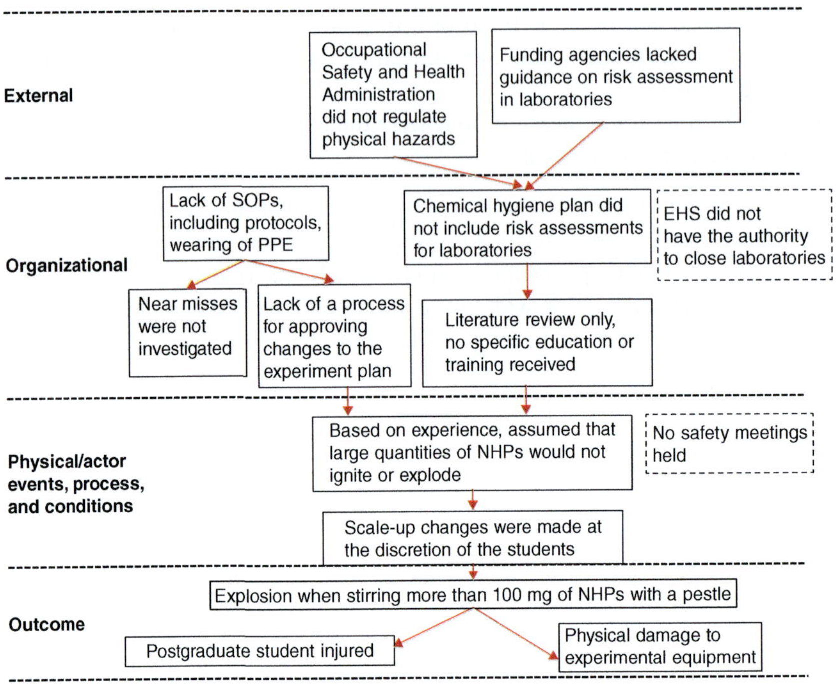


**Figure 3.6** Analysis of accidents at University A using AcciMap. *Note:* Red arrows indicate causal relationships. The dashed box represents accident factors for which no causal relationship can be found anywhere.

#### 3.3.4.1 Analysis of the Accident at University A Using the AcciMap

As an alternative to STAMP, the analysis was carried out in AcciMap, which is suitable for analyzing accidents at University A and incorporates systems thinking. The safety recommendations to be taken by the higher level are logical and clear, with arrows showing the causal factors that led to the accident. However, it should be noted that not all accident factors are visualized and that there are some factors that are not causally related. The fact that the EHS, which carries out workplace inspections, is not authorized to close the laboratory if it detects an action or situation that could lead to a high-risk accident, shown by the dotted line in the figure, is an important safety factor, but the causal link is not clear. At University A, two near misses had occurred prior to the accident, but no accident investigation or analysis had been carried out. Given the importance of the near-miss investigations listed in Chapter 2, it is not clear how this fact is reflected in the safety recommendations (Figure 3.6).

## 3.4 Accident Models Applicable to Each Industry

### 3.4.1 Characteristics of Each Industry

Perrow (1984) illustrated the characteristics of each industry on a coordinate diagram, with the degree of coupling between systems and components on the vertical axis and whether the interaction of systems or processes is linear or complex on the horizontal axis. A collection of units, parts, and subsystems is defined as a component, and a system is defined as a collection of components. According to Perrow (1984), the characteristics of each industry are classified according to the following indicators.

Tightly coupled systems, synonymous with the engineering term coupling, have no looseness, the design of the entire process is the only way to achieve the objective, and the sequence of the process is invariant and time dependent. However, in a loosely coupled system, there are many ways to achieve the goal: substitutes can be used for the units, parts, etc. that make up the system; the sequence can be changed; and process delays can be accommodated flexibly. For example, in nuclear power plants, part of the former tightly coupled system, fuel cannot be substituted from nuclear to other materials, whereas in thermal power plants, oil can be substituted for coal or coal for oil. In manufacturing plants such as car factories, which belong to the latter loosely coupled system category, if a problem occurs in the production of a door or a seat on the assembly line, it can be replaced later with a component that meets the specifications to produce a finished product, whereas in nuclear power plants and chemical plants, it cannot be replaced later.

Linear interaction allows people to see and predict what is happening in the process upstream and downstream of the faulty part, even when a unit or other part is malfunctioning, because these interactions are predictable and obvious, regardless of the number of units and parts. In a linear system, in a manufacturing plant, worker errors and interactions can be confirmed, so if a worker misunderstands the instructions and does the work, the supervisor can see the error and identify it.

In complex interactions where components have multiple functions, either intentionally or unintentionally at the design stage, it is not possible to see what is happening in the process when a malfunction occurs. As a result, the full extent of the situation cannot be determined immediately after a fault occurs, as there is nothing to do but guess at the situation. In such complex systems, not only do failures reveal unexpected interdependencies between components, but the specialization of the roles and knowledge of the people working in them means that they cannot predict, notice, or analyze these interdependencies.

### 3.4.2 Industry Characteristics and Accident Models

Hollnagel and Speziali (2008) modified the interaction of the horizontal axis defined by Perrow by changing and removing some of the industries classified in each quadrant. The horizontal axis was changed to the difficulty of tracing or tractability, which indicates whether the system is easy to describe or not. Traceable or tractable means that the operating principle of the system or process is known and simple and that the description and the system match. Intractable means that some or all of the functions of the system or process are unknown, the representation is dense, and the system may be changed before the representation is complete. In addition to the changes to the horizontal axis, modifications were made to remove continuous processing, multipurpose government agencies, nuclear weapons accidents, and genetic modification from the industry diagram to show the range of applicability of each accident model.

According to Hollnagel and Speziali's classification (2008), the sequential accident model is applicable to manufacturing industries with assembly lines, most manufacturing industries, vocational schools, junior colleges, post offices, and mining, which are classified in the third quadrant. The sequential accident model is only applicable to industries where the cause and effect are clear, i.e. the interaction between the components is linear, and industries where the interaction is complex become unanalyzable because the factors leading to the accident cannot be traced. The sequential accident model is therefore not applicable to industries where the interactions are complex.

The epidemiological accident model can be applied to dams, power plants, maritime transport, rail transport, and military early warning, which are classified in the first quadrant. The Late SCM, an epidemiological accident model, can be applied to industries where linear elements are excluded and interactions are complex, but the link between cause and effect limits its application to socio-technical systems in general.

The systemic accident model fits the second quadrant industries of nuclear power plants, chemical plants, aerospace, financial markets, aircraft, and airports.

Fukuoka and Furusho (2022) analyzed accidents in universities and published that the epidemiological accident model is applicable. Therefore, further modifications to Hollnagel and Speziali's industrial application diagram are shown in Figure 3.7.

Hollnagel and Speziali (2008) explained the reasons for the change in accident models by stating that the working environment of people had changed from production activities using simple machines to production activities using complex systems with computer technology during the period from the industrial revolution to modern socio-technical systems. The nature of accidents has also changed in line with these changes, from simple accidents with clear cause and effect and few underlying factors to complex accidents with unclear cause and effect and many underlying factors. Accident models have also evolved to adapt to these changes, moving from the third quadrant to the first and then to the second quadrant in Figure 3.7.

One of the requirements for selecting an appropriate accident model for each industry is to identify the specifics of the industry to which the reader's organization belongs, i.e. the difficulty of tracking and coupling, given the historical background and industry specifics of the aforementioned accident models, and to select an applicable accident model.

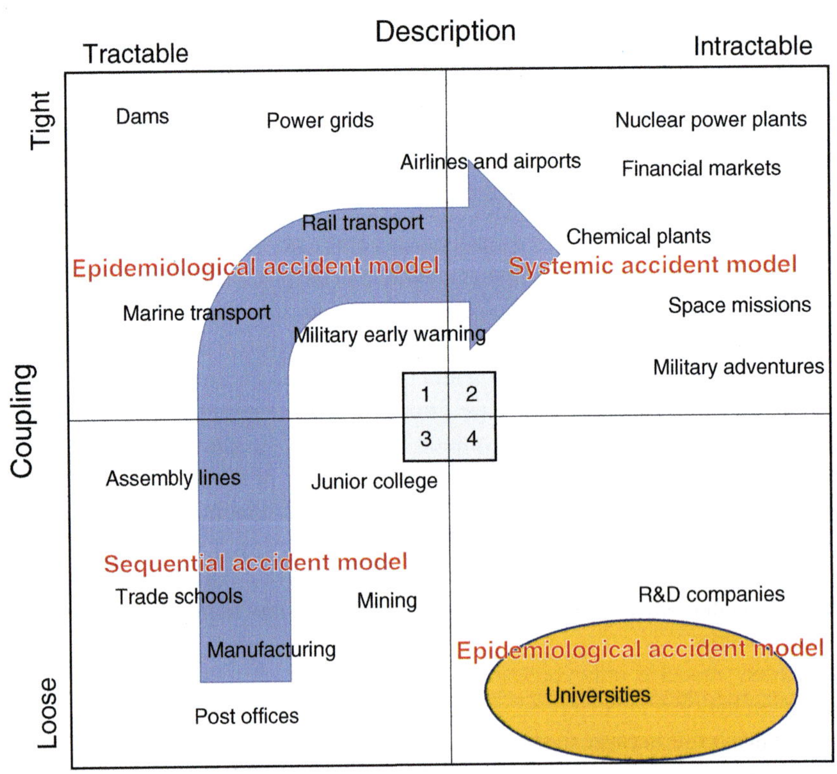


**Figure 3.7** Accident models applicable to different industries. *Source:* Adapted from Fukuoka (2017). *Note:* The arrow from the third quadrant to the second quadrant indicates the direction of development of industrial systems and accident models. The orange color in the fourth quadrant indicates universities, where no applicable accident models were identified by 2022, but epidemiological accident models were found to be applicable (Fukuoka and Furusho 2022).

Fukuoka (2017), as an accident investigator for the Japan Transport Safety Board, has analyzed a large number of marine accidents that occurred in Japan as well as marine accidents that occurred worldwide as an analyst for the correspondence group of the subcommittee on implementation of IMO instruments. In analyzing these accidents, the accident models applicable to shipping were epidemiological accident models, as mentioned above, but ship engine failures were similar to those in the third quadrant manufacturing industry, so it was appropriate to use sequential accident models, such as variation tree analysis (VTA). Marine accidents, such as collisions, groundings, fires, capsizes, and casualties, occurred in complex socio-technical systems and used Late SCM, which was appropriate for industrial accident analysis in the first quadrant.

It would be useful for accident investigation and analysis personnel in each industry to learn the analysis methods of each accident model and to use different accident models depending on the form of the accident, in addition to the characteristics of the industry to which the university or company belongs.

## 3.5 Conclusion

The classic accident model is the domino theory, in which events occur in a linear chain of dominoes, and many accidents have been investigated and analyzed using this model, and measures have been taken to prevent the recurrence of accidents. In modern society, where people work in close contact with society and a wide variety of systems, the environment surrounding people has become more complex and socio-technical systems have developed. Even using models that classify accidents that occur in complex socio-technical systems as sequential accident models, it was no longer possible to get the full picture of the factors that cause accidents. Accident models were developed in the 1970s and 1980s through the study and analysis of the various serious accidents that occurred in many industries, and there are currently three main categories of accident models.

In this chapter, the University A explosion accident is reanalyzed using five different types of accident models from the three broad categories of sequential accident models, epidemiological accident models, and systemic accident models in order to clarify the characteristics and problems of each type of accident model.

First, the domino theory, represented as a sequential accident model, fails to derive many other accident-contributing factors, including the operation of SMSs. Even if an accident is investigated and analyzed using the domino theory and corrective action is taken to address the problems based on the results, many accident-contributing factors will not be visualized and will remain latent. Therefore, it is clear that if an accident to which the sequential accident model is inherently inapplicable is investigated and analyzed using the domino theory approach, even if accident-contributing factors are identified, the safety measures derived from them will be neutralized and the accident will recur.

In the reanalysis of the explosion accident at University A using HFACS, which is part of the epidemiological accident model, there are no subcategories corresponding to the involvement of external organizations affecting the safety of the university, and there are accident factors that cannot be extracted using the HFACS classification method.

The analysis using the SHEL–Reason hybrid model has the advantage of providing a direction for planning safety recommendations, as the unsafe acts are designed to specify the type of error and violation. However, the structure of the model does not allow the same issue as in HFACS to be derived to develop accident-contributing factors and safety recommendations beyond the organization of the operator who makes an error.

The ATSB model consists of five levels: (1) organizational influence, (2) risk control, (3) local conditions, (4) individual action, and (5) occurrence event. The advantage is that the actions to prevent recurrence are clear, as the contributing factors that fall under organizational influence and risk control are defined as safety issues, and these safety issues are the subject of safety recommendations. A weakness is the complexity of the work process, as new analysis tools are used to extract the factors at each level and a relatively long process is required to identify the safety factors. Regarding accidents that have occurred, the ATSB model raised some interesting points. The terms cause, root cause, and proximal cause are not used for the purpose of safety investigation, because the more complex an accident is, the more difficult it is to identify, even after a thorough investigation, and because cause is a term relating to responsibility and duty. Root cause and proximal cause are simplified and replaced by contributing safety factors.

Analysis with the RMQMP model clearly shows a relationship between the hole and latent conditions that have not been elucidated by other epidemiological accident models. Although the causal relationship of each accident-contributing factor is not shown, the logical identification of the location of the hole in the system and all the accident-contributing factors required to rectify the hole is a major advantage. The approach to accident prevention and proactive prevention in the model is the construction of multiple layers of defense where a single error does not lead directly to a serious accident; the model states that accidents can be prevented by correcting the higher level accident-contributing factors contained in each element of the SHEL model.

Finally, in the analysis in AcciMap using systems theory, the safety recommendations to be taken by the higher level are logical and clear, with arrows depicting the causal factors that led to the accident. However, there are weaknesses: not all accident factors are visualized, there are factors with no causal relationship, and it is unclear how important accident factors are reflected in the safety recommendations.

Each accident model has a sector or type of accident to which it can best be applied due to the historical context in which the model was developed. The reader is advised to identify which sector their organization belongs to and use the appropriate accident model. In doing so, it is also necessary to select an accident model according to the type of accident, such as a sequential accident model for the investigation and analysis of machine failure, or an epidemiological accident model for the investigation and analysis of fire, explosion, and casualty accidents.

## References

Australian Transport Safety Bureau (2008). ATSB Transport Safety Research Report: Aviation Research and Analysis Report-AR-2007-053, Analysis, Casualty and Proof in Safety Investigations.

Davis, J., Davies, J.M., and Flemons, W. (2007). The Calgary health region: transforming the management of safety. In: *Multimodal Safety Management and Human Factors* (ed. J. Anca), 31–40. Hampshire: Ashgate Publishing Limited.

Dekker, S. (2006). *The Field Guide to Understanding Human Error*. Surrey: Ashgate Publishing Limited.

Dell, G. (2007). Aerial Agriculture Accidents 2000–2005: The human factors and system safety lessons. In: *Multimodal Safety Management and Human Factors* (ed. J. Anca), 113–129. Hampshire: Ashgate Publishing Limited.

Endsley, M.R. (2000). Theoretical understandings of situation awareness: a critical review. In: *Situation Awareness Analysis and Measurement* (eds. M.R. Endsley and D.J. Garland), 3–32. Mahwah: Lawrence Erlbaum Associates, Inc.

Fukuoka, K. (2016a). Visualization of a hole and accident preventive measures based on the Swiss cheese model developed by risk management and process approach. *WMU Journal of Maritime Affairs* 15 (1): 127–142.

Fukuoka, K. (2016b). Model of marine accidents developed by the Swiss cheese model and recommendations for systematic accident prevention. Doctoral thesis. Kobe University (Japanese).

Fukuoka, K. (2017). Accident models and their applicable industries. *Journal Science of Labour* 93 (2): 48–60. (Japanese).

Fukuoka, K. (2019). *Safer Seas: Systematic Accident Prevention.* Boca Raton: CRC Press Taylor & Francis Group.

Fukuoka, K. and Furusho, M. (2016). Relationship between latent conditions and the characteristics of holes in marine accidents based on the Swiss cheese model. *WMU Journal of Maritime Affairs* 15 (2): 267–292.

Fukuoka, K. and Furusho, M. (2022). A new approach for explosion accident prevention in chemical research laboratories at universities. *Scientific Reports* 12 (1): 3185.

Grech, M., Horberry, T., and Koester, T. (2008). *Human Factors in the Maritime Domain.* Boca Raton: Taylor & Francis Group.

Heinrich, H.W. (1959). *Industrial Accident Prevention.* New York: McGraw-Hill Book Company, Inc.

Hollnagel, E. (2004). *Barriers and Accident Prevention.* Surrey: Ashgate Publishing Limited.

Hollnagel, E. (2013). *Safety Analysis of Socio-technical Systems FRAM Guidebook* (trans. Komatsubara A). Tokyo: Kaibundo.

Hollnagel, E. and Speziali, J. (2008). Study on Developments in Accident Investigation Methods: A Survey of the "State-of-the-Art," SKI Report 2008:50. Stockholm: Swedish Nuclear Power Inspectorate.

International Civil Aviation Organization (1993). Human Factors Digest No.7: Investigation of Human Factors in Accidents and Incidents; Circular 240-AN/144. Montreal: International Civil Aviation Organization.

International Maritime Organization (2000). Amendments to the Code for the Investigation of Marine Casualties and Incidents: The IMO/ILO Process For Investigating Human Factors; Resolution A. 884(21).

International Maritime Organization (2014). *Safety Investigation Into Marine Casualties And Marine Incidents*, Model course 3.11, 2014 ed.

International Maritime Organization (2015). Marine Casualty Investigation Investigators' in-the-Field Job aid MAIIF/IMO; Sub-Committee on Implementation of IMO Instruments (III), Annex 4.

Japan Marine Accident Inquiry Agency (2008). *IMO Marine Casualty Investigator's Manual* (English-Japanese Bilingual). Tokyo: Kaibundo.

Landre, J.D., Irving, M., Hodges, I. et al. (2007). Learning from accidents and incidents. In: *Multimodal Safety Management and Human Factors* (ed. J. Anca), 131–142. Hampshire: Ashgate Publishing Limited.

Leveson, N.G. (2011). *Engineering a Safer World: Systems Thinking Applied to Safety.* London: The MIT Press.

Marine Accident Investigators' International Forum (2015). MAIIF investigation manual. http://www.maiif.org/images/MAIIF%20Manual%202014.pdf (accessed 13 December 2015).

McNair, D. (2007). Unsafe or safety-significant act? In: *Multimodal Safety Management and Human Factors* (ed. J. Anca), 17–30. Hampshire: Ashgate Publishing Limited.

Perrow, C. (1984). *Normal Accidents.* Princeton: Princeton University Press.

Rasmussen, J. (1997). Risk management in a dynamic society: a modelling problem. *Safety Science* 27 (2–3): 183–213.

Reason, J. (1990). *Human Error*. Cambridge: Cambridge University Press.

Reason, J. (1995). A systems approach to organizational error. *Ergonomics* 38 (8): 1708–1721.

Reason, J. (1997). *Managing the Risks of Organizational Accidents*. Surrey: Ashgate Publishing Limited.

Reason, J. (2008). *The Human Contribution: Unsafe Acts, Accidents and Heroic Recoveries*. Surrey: Ashgate Publishing Limited.

Reason, J., Hollnagel, E., and Paries, J. (2006). Revisiting the Swiss Cheese Model of Accidents. Bretigny-sur-Orge: EUROCONTROL Experimental Centre.

Salmon, P.M., Regan, M., and Johnston, I. (2007). Managing road user error in Australia. In: *Multimodal Safety Management and Human Factors* (ed. J. Anca), 143–156. Hampshire: Ashgate Publishing Limited.

Schröder-Hinrichs, J.U., Baldauf, M., and Ghirxi, K.T. (2011). Accident investigation reporting deficiencies related to organizational factors in machinery space fires and explosions. *Accident Analysis and Prevention* 43 (3): 1187–1196.

Underwood, P. and Waterson, P. (2013). System thinking, the Swiss Cheese Model and accident analysis: A comparative systemic analysis of the Grayrigg train derailment using the ATSB, AcciMap and STAMP models. *Accident Analysis and Prevention*. 68: 75–94.

Wiegmann, D.A. and Shappell, S.A. (2001). *A Human Error Analysis of Commercial Aviation Accidents Using the Human Factors Analysis and Classification System (HFACS)*, DOT/FAA/AM-01/3. Washington, D.C: Federal Aviation Administration.

Wiegmann, D.A. and Shappell, S.A. (2003). *A Human Error Approach to Aviation Accident Analysis*. Surrey: Ashgate Publishing Limited.

# 4

# Risk Management and Safety Management Systems

## 4.1 Introduction

The National Academy of Sciences (2011) states as follows:

*Many features of effective EHS management are identical to management practices advocated by proponents of quality assurance and business excellence... Within many organizations, some elements of EHS management are already in place, such as policy and risk assessment records, but other aspects need to be developed. It is important that all the elements described here are incorporated into the EHS management system.*

Investigations and analyses of accidents that have occurred at universities and academic institutions, as well as reanalyses of accident investigation reports, have revealed that accidents have occurred and faculty, researchers, and students have become victims due to inadequate risk assessments, such as the identification of hazards and risk analyses, and insufficient SMSs in laboratories and universities, regardless of whether the accidents occurred in chemical laboratories or during fieldwork.

This chapter looks at the risk management processes and SMSs that are needed to prevent accidents and mitigate the damage that occurs in universities, etc. at an organizational level. These are also the theories that need to be understood in order to carry out scientific accident investigations in the event of an accident. In universities, risk assessment is taken up as a subject to be implemented as part of education and training in laboratory safety, and the need for it is recognized. However, risk management processes and SMSs, in particular, are completely new concepts for universities, and their need is not fully recognized.

Following the accident involving an explosion during a chemical experiment at University A, the US Chemical Safety Board (CSB 2010), the Occupational Safety and Health Administration (OSHA 2013), and others expressed the need to introduce a modified SMS to fit the actual conditions of universities, etc. rather than introducing the SMS as it stood.

This chapter discusses a risk management process that extends the framework of risk assessment to reflect the reality of organizational structures and accidents during experiments and field research activities and describes a modified SMS that has proved useful in preventing accidents and that can be applied to universities, etc.

*Accident Prevention and Investigation: A Systematic Guide for Professionals, Educators, Researchers, and Students*, First Edition. Koji Fukuoka.

## 4.2 Differences Between Risk Assessment and Risk Management

Risk assessment is recognized and implemented as part of laboratory safety. However, as indicated at the beginning of Chapter 1, there has not been a paradigm shift in laboratory safety due to the diverse and numerous accidents that occur in universities, etc. Looking at other industries, a process has been added to check whether the effects of the risk assessment have been realized after it has been carried out. In other words, if the risk reduction is ineffective or a new risk arises, the work is stopped and the risk assessment is repeated, for example by investigating and analyzing the causes of the ineffectiveness and then correcting the deficiencies found.

The results of the accident investigations and analyses of the university accidents show that unusual events or situations occurred during experiments with chemicals or high-pressure gases in the laboratory or during scuba diving in field research activities at sea, but these events or situations were not recognized as invalidating the effectiveness of the risk assessment or creating new risks. As a result, postgraduate students and researchers became victims due to the continuation of research. In the case of the chemical explosion accident that occurred at University A, if the two near misses that occurred prior to the accident had been properly monitored and reviewed, the factors that caused this explosion could have been identified in advance, and there would have been an opportunity to correct the unsafe conditions of the factors and prevent the accident.

The difference between risk assessment and the risk management process is as follows. The risk assessment reduces the risk to an acceptable level based on the principles of risk reduction when the risk assessment shows that the risk is not acceptable. The risk management process includes monitoring and review process whether the risk events reduced by the risk assessment are controlled during operation and, if not, new measures are taken to control the risk. The monitoring and review process may be carried out by the individual researcher conducting the experiment, a colleague, a supervisor, or in the case of offshore field activities, a diving safety officer.

The main differences between the steps in the risk assessment and risk management processes are shown in Table 4.1. The terminology and process for risk assessment is based on ISO/IEC Guide 51 (ISO/IEC 1999) and the risk management process is based on IEC/ISO 31010 (2009). Note that in the risk management process, hazard identification is replaced by risk identification.

**Table 4.1** Key differences between risk assessment and risk management processes.

| Risk assessment | Risk management process | Description of the risk management process |
|---|---|---|
| Hazard identification | Hazard identification | Identify the causes and sources of the risk |
| Risk estimation | Risk analysis | Determine the consequences and probabilities of the risk |
| Risk evaluation | Risk evaluation | Estimate the level of risk |
| Tolerable risk achieved if not repeat risk reduction | Risk treatment | Decide whether and how to treat the risk |
| | Monitoring and review | Monitor the situation that could change/ invalidate the risk assessment, and if necessary, risk assessment can be updated |
| Iterative process of risk assessment and risk reduction | Iterative process of risk management | |

However, as hazard identification is used in safety and in laboratory safety risk assessments, the term hazard identification will be used in this book.

## 4.3 Concept of the Risk Management Process

According to IEC/ISO 31010 (2009), the risk management process includes hazard identification, risk analysis, risk evaluation, risk treatment, and monitoring and review. Hazard identification is the identification of causes or sources of risk, the nature of the event or situation, and its consequences. Risk analysis is the determination of the level of risk by measuring the consequences of the identified risk events and the probabilities of the risk events occurring.

The risk management process described in this chapter is a series of actions taken by the faculty, principal investigators (PIs), and students in the laboratory to ensure that experiments are safe by following a process before and during the experiment.

It is important to understand the meaning of the terms hazard and risk. A hazard is a potential source of harm, and in the field of safety, generally, it refers to an undesirable and dangerous event or condition. It is often used with event as a preposition, e.g. a fire hazard, an explosion hazard, a toxic hazard, and a drowning hazard. Risk, however, refers to the effects caused by a hazard and consists of the severity and frequency of the expected damage. In other words, the level of risk can be expressed as a measurable number using severity and frequency.

According to the National Academy of Sciences (2011), risks and hazards are distinguished as follows. A hazard is a danger inherent in a particular chemical substance or operation, and therefore, the hazard of that chemical substance cannot be reduced. The hazard also cannot be reduced by the process or operation in which the chemical is handled in an experiment. Risk refers to the probability of injury resulting from handling a particular chemical substance or the probability of injury occurring in the course of a series of experiments to achieve a particular result with that substance. Therefore, the risk associated with this chemical substance can be reduced.

Real-life examples will be used to illustrate this. For researchers and students conducting field research in Okinawa, Japan, the bite of a venomous snake, such as the habu (scientific name: *Protobothrops flavoviridis*), is a threat that can cause death from the toxins in the venomous snake's saliva in areas where there is no adequate medical system. In Okinawa prefecture in southern Japan, there are currently around 100 cases of habu attacks per year; nowadays, with the development of medical systems and palliative care, there are almost no deaths. When attacking, the habu shrinks its body into an S-shape and then quickly extends its body to bite the target. They do not jump, and two-thirds of the habu's total length is the maximum distance they can attack (Okinawa Prefectural Government 2024). Therefore, if a person is generally present at a distance greater than the length of the hub, it is physically impossible to inflict a bite wound on a person in a single attack. This means that the risk of injury from a hub attack is reduced. Okinawa Prefecture recommends that people keep a distance of 1.5 m or more if they find a hub while working.

In experiments with nickel hydrazine perchlorate (NHP), which is explosive as an inherent property of the chemical, scale-up means that there is an explosion hazard. However, scale-up in milligram is in itself a risk-reducing measure because even if the substance were to explode, the severity of the explosion to humans would be low. Other risk reduction measures for explosive substances include the provision of equipment for rapid cooling of the reactants in the event of a runaway reaction and the addition of additives to inhibit the reaction and taking these measures in the event of a runaway reaction.

### 4.3.1 Hazard Identification Methods

The first process in ensuring the safety of educational and research activities is hazard identification. The nature and characteristics of hazards will vary according to the area being studied, but the following are some of the areas where accidents are particularly common in universities, etc.

- In the field of chemicals, the hazards of each substance are listed in a safety data sheet (SDS) provided by the manufacturer when the chemical is purchased. SDSs are applicable to industrial organizations that use large quantities of chemicals, but some topics are not suitable for universities, etc. that use small quantities of chemicals.
- Laboratory chemical safety summaries provide specific data on chemicals that are most commonly used in laboratories. Although fewer chemicals are listed than in SDSs, the data include physical, chemical, and toxicological data to allow relative comparisons of hazards and to facilitate risk assessment by researchers (National Academy of Sciences 2011).
- Hazard information for each chemical is available on the websites of the occupational health and safety agencies in many countries.
- Hazards arising from the use of equipment and instruments in chemical experiments are operational hazards. Operational hazards can be related to the prohibitions listed for each process in the experimental plan and in the equipment operating manuals, but if detailed accident investigation reports are available, it is possible to identify the process in which the accident occurred, and the operational hazard is identified. Publication of these detailed accident investigation reports is limited and not widely available. The accident investigation reports on chemical substances published by CSB in the United States are based on evidence-based investigations and analysis, so their content can be used as a reference for hazard identification.

For field research activities, especially in the sea and mountains, the following applies:

- When conducting research activities at sea, check the website to see whether the experimental area is inhabited by dangerous marine species, such as stingrays, moray eels, sea snakes, stonefish, Portuguese man-of-wars (*Physalia physalis*), and great white sharks.
- When diving or snorkeling, information on strong currents, strong tidal streams, and seabed topography may be available from national hydrographic agency or coast guard agency websites. It should be noted that currents that differ from the direction of surface currents may exist in the middle or deeper levels.

  In waters where swimmers have been swept out to sea by rip currents and where drownings have occurred in the past, details about the accidents can be found out on websites such as the coast guard's website, depending on the country. The mechanism of rip currents is as follows. Waves wash ashore from offshore. Seawater accumulates on the shore and, as a result, seawater that has nowhere else to go flows toward the offshore area. This fast flow is called a rip current. According to the Japan Coast Guard, rip currents are 10–30 m wide and about 200 m long, with a maximum length of 300 m in Japan. They can occur anywhere along the coast. These beaches can be found on websites such as the coast guard, depending on the country where the accidents have occurred. Notice boards put on beaches can provide important information about the risk of drowning due to rip currents. The information on these boards is a warning to researchers and students from local authorities and administrations to prevent a recurrence, as many people have died in the same area. If you are caught in a rip current, wave your arms from side to side for help, do not swim against the current, let it flow first, and then swim parallel to the shore. Once out of the rip current, swim toward the shore.

- Climate change has increased the likelihood of being involved in natural disasters during field research. Hazards from natural disasters, including mudslides, floods, and landslides, can be found on hazard maps published on the websites of national and local authorities, meteorological services, and national river management agencies. These hazard maps include inundation zones for floods and river bursts, inundation depths, and evacuation points. It will be necessary to check the weather forecasts issued by the Meteorological Office frequently and to consider changing the dates and content of the experimental plans when weather warnings are issued.
- Some areas are prone to river flooding due to torrential rainfall. These are areas that have been frequently flooded by rivers in the past and areas that used to be rivers but have been converted to flat land or housing through land reclamation or development. Because these areas were originally rivers and lowlands, they are more susceptible to flooding from rivers and from water flowing in from surrounding areas during torrential rainfall. In some countries, maps showing former river areas can be found on the website of the geospatial information authority.
- The best way to learn about operational hazards is to carefully examine comprehensive accident investigation reports published by government agencies, universities, or private organizations. For example, details of personal accidents involving ships, such as research vessels and diving boats, can be obtained from accident investigation reports published on the websites of accident investigation organizations in each country that is a member of the International Maritime Organization (IMO). Lessons can be learned from accident investigation reports, such as the hazards inherent in the planning of dives and experiments and the methods that can be used to prevent accidents. In a small number of universities, in the hope of preventing a recurrence of the same accident, a third-party accident investigation committee has been set up to investigate the accident that occurred at the university, and the results have been made public in the form of an accident investigation report.
- Hazards in the mountains include getting lost, hypothermia in bad weather, freezing to death, slipping, being caught in an avalanche, etc. A device used exclusively in the mountains is useful to prevent getting lost. These devices use the Global Positioning System to record your position on mountain routes approved by the local authorities. It is advisable to check the websites of the local authorities responsible for mountain areas, as they sometimes publish on their websites the dangerous areas where slips, falls, and other accidents have occurred in the past.
- For the latest information on hazards caused by natural phenomena, such as hurricanes, typhoons, and tornadoes that are unique to the region, visit the websites of the meteorological agencies and organizations involved in natural disaster prevention in each country. For typhoons in East Asia, the Japan Meteorological Agency's website provides forecasts of the typhoon's path three to four days in advance, as well as storm zones where average wind speeds generally exceed 25 $m\,s^{-1}$. Windy.com provides weather information on typhoons, hurricanes, and other natural phenomena around the world, including forecasts for the next few days or longer.

### 4.3.2 Risk Analysis Methods

There are three methods of risk analysis: qualitative, semiquantitative, and quantitative, according to IEC/ISO 31010 (2009).

Qualitative methods are a way of classifying risk events using adjectives such as very large, large, medium, small, and very small as severity scales and expressing them as frequency scales such as frequent, probable, occasional, remote, and unlikely. It is used when objective data on the event in question over a long period of time are not available and expert opinion should be consulted.

Quantitative methods are methods in which severities and frequencies are expressed using numbers in the light of historical data, statistics, etc. without the use of adjectives such as those used in qualitative methods. Historical data on the accident or event must be accumulated and reliable. Quantitative methods are considered undesirable because of insufficient information about the system or event being analyzed, insufficient data, and many uncertainties, such as the influence of human factors, which cannot be expressed numerically.

A semiquantitative method is a method that combines qualitative and quantitative methods, using reliable numerical data obtained on severity or frequency scales and using terms from qualitative methods for scales where data are lacking. Semiquantitative and qualitative methods are commonly used as risk analysis methods.

Reanalysis of CSB accident investigation reports or university accident investigation and analysis reports shows that the parties involved in these accidents did not carry out proper risk assessments and judge the level of risk based on their own past successful experience. Relying on experience can lead to inadequate or inappropriate risk control. To avoid this, proper risk analysis and assessment is required. The following are some of the methods that are useful for universities, etc. when carrying out risk analysis or preventing accidents.

Severity scales and frequency scales are used in risk analysis. There are several types of scales, including nominal and ordinal scales. Maguire (2006) provides the following scales, which are very clear and easy to measure, despite some problems. Table 4.2 shows the scales used in some industries. The scale shows a relationship where one death equals 10 serious injuries, and this ratio is controversial. Table 4.3 presents frequency scales, leaving some ambiguity about "probable" and "occasional." Table 4.4 provides a clearer definition of frequency scales, but is not considered suitable for industries requiring more detailed scales or scales longer than 100 years (Fukuoka 2019).

### 4.3.3 Risk Evaluation Method

Risk evaluation is the process of determining whether the level of risk estimated in the risk analysis is an acceptable level for the laboratory or organization. A risk matrix combining severity scales and frequency scales is used in risk evaluation. The risk matrix indicates whether the identified risk event is acceptable to your organization, and this decision should be discussed and agreed upon within your organization. In the case of universities, this would need to be agreed upon by the risk management committee.

**Table 4.2** Severity scales.

| Nominal scale | Ordinal description |
|---|---|
| Cataclysm | Thousands of fatalities and injuries |
| Disaster | Hundreds of fatalities and injuries |
| Calamity | Tens of fatalities |
| Catastrophic | Single figure, multiple deaths (up to 10) |
| Critical | Single death and/or multiple severe injuries |
| Severe | Single or several serious injuries |
| Significant | Several minor injuries – significant loss of time |
| Minor | Single minor injury |

**Table 4.3** Frequency scales (1).

| Nominal scale | Ordinal description |
|---|---|
| Frequent | Likely to be continually experienced |
| Probable | Likely to occur often |
| Occasional | Likely to occur several times |
| Remote | Likely to occur sometime |
| Improbable | Unlikely, but may exceptionally occur |
| Incredible | Extremely unlikely |

**Table 4.4** Frequency scales (2).

| Nominal scale | Ordinal description |
|---|---|
| Frequent | Weekly |
| Probable | Monthly |
| Occasional | Annually |
| Remote | 10 yearly |
| Improbable | 100 yearly |

In many industries, including the aviation and chemical industries, safety requirements and demands on organizations have changed from lenient to more stringent as safety technology has developed. In other words, the criteria for what is acceptable or unacceptable for an organization, such as a university, may change over time. It is therefore prudent for organizations to keep abreast of changes not only within their own organization but also in their environment and to be sensitive to the trends and directions in safety at any given time.

According to IEC/ISO 31010 (2009), risk levels can be divided into four categories.

1. Risks cannot be justified except in exceptional circumstances.
2. Risks are acceptable only if the benefits substantially outweigh them.
3. Risks are acceptable if the benefits outweigh the risks.
4. Risks are widely accepted.

The unacceptable region is (1), the as low as reasonably practicable (ALARP) or acceptable regions are (2) and (3), and the widely acceptable region is (4).

The risk matrix uses this ALARP concept to collate and tabulate the four categories (1) to (4), each with a risk level from A to D. Table 4.5 shows a sample risk matrix available to universities, etc. based on Tables 4.2 and 4.4, where A corresponds to (1) and the risk is not justified, D corresponds to (4) and the risk is widely accepted, etc.

The risk matrix for other industries is heavily influenced by the pursuit of profit as one of the objectives of the organization, and unlike Table 4.5, there are fewer A categories and more B and C categories. B and C categories would mean that risk reduction measures can be taken and operations can begin. However, universities, etc. unlike businesses where the pursuit of profit is the top priority, must take into account the exploration of frontiers, the development of cutting-edge technology, advanced education, and the safety of staff and students in

**Table 4.5** Risk matrix.

| | Calamity | Catastrophic | Critical | Severe | Minor |
|---|---|---|---|---|---|
| Frequent | A | A | A | B | B |
| Probable | A | A | A | B | C |
| Occasional | A | A | A | C | D |
| Remote | A | B | B | D | D |
| Improbable | B | C | C | D | D |

educational and research activities. It is considered that high-risk events in laboratory experiments and field research activities should be operated at category C or D in the risk matrix, with sufficient risk reduction in accordance with the hierarchy of control.

### 4.3.4 Risk Treatment

Risk treatment is the selection and implementation of options that can change the severity or frequency, or both, of risk events. The hierarchy of control must also be applied here. For example, in chemical experiments carried out at universities, etc., the use of chemicals that are expected to cause a fire explosion is limited from the outset to a small amount of milligrams, and a level of risk is allowed such that even if an explosion or other event occurs, it will not cause human casualties. This approach is included in risk treatment. With regard to natural disasters, the frequency of earthquakes, typhoons, and hurricanes cannot be artificially altered by people through the use of science. However, the impact or severity of these natural disasters on people and property can be changed by harnessing the power of science or by taking steps to mitigate the damage in advance. In the case of earthquakes, human casualties and property damage can be reduced by designing buildings to higher seismic resistance standards, by moving and securing heavy equipment and machines in laboratories at high elevations to lower locations, and by ensuring that faculty, staff, and students are trained in life-saving actions after an earthquake.

### 4.3.5 Monitoring and Review

Monitoring and review is the periodic inspection of the operational situation to ensure that the risk controls adopted through the risk assessment process remain effective or that no new risks have arisen. If necessary, the results of the inspection should be used to carry out another risk assessment to reduce the risk and resume the operation.

In an accident at a university where an explosion occurred while handling NHP, two near misses had occurred in the same laboratory prior to the accident. However, these near misses were not shared with the whole laboratory personnel and were not investigated. The fact that the near misses occurred before the accident was a fact that the previous risk assessment was no longer valid, and this fact was expressed as an event or near miss that people could recognize.

The monitoring and review process involves a comprehensive examination of the procedures, equipment, etc. currently in use for any problems, and if any problems are found, they are to be rectified, and then the experiment is carefully repeated with revised procedures.

The monitoring and review process in other industries and in the maritime industry is as follows. When a cargo ship arrives at the quay for loading and unloading, before the crew and workers enter an enclosed space, such as a cargo hold or pump room, it is required that the concentration of the atmosphere be measured at several heights from the floor in the space using an oxygen concentration detector. If the measured oxygen concentration is below 21% and within the unacceptable risk level range, the oxygen concentration must be increased by forced ventilation. Only when the oxygen concentration has increased to 21% may the person enter the cargo hold and begin work. During the work, the oxygen concentration must be measured regularly to ensure that the atmosphere in the cargo hold is within safe limits. If the oxygen concentration drops during the work, the work must be stopped, the persons evacuated to a safe area, and the forced ventilation restarted.

The procedures used in other industries, as described in the following procedures incorporating risk management processes, are considered to be helpful for the safety of laboratories at universities, etc., as the risk management processes are embedded as part of the operational system.

### 4.3.6 Procedures Incorporating Risk Management Processes

In the shipping industry, the Convention on the International Regulations for Preventing Collisions at Sea, 1972 (COLREGs), which is the international rules for avoiding collisions on ships developed by the IMO, procedures for developing voyage route plans (IMO 2000), and procedures for operations on enclosed spaces (IMO 2011), meaning work in enclosed areas, such as cargo tanks and cargo holds, have risk management processes incorporated.

The following is a specific example of how international regulations to prevent collisions between ships at sea incorporate a risk management process (Fukuoka 2019). Collision avoidance procedures are stipulated in COLREGs and compared to the risk management process based on ISO 31000 and IEC/ISO 31010 (ISO 2009, IEC 2009).

Rule 5 covers risk identification; Rules 7(a), (b), and (c) cover risk analysis; Rule 7(d) covers risk evaluation; Rules 8(a) and (b) cover risk treatment; and Rule 8(d) covers monitoring and review. In the case of restricted visibility, the first paragraph of Rule 19(d) covers risk analysis, and Rules 19(d)(i) and (ii) and 19(e) cover risk treatment.

The pertinent sections of COLREGs are as follows:

1. Rule 5 states that every vessel must always maintain a proper lookout to enable a full appraisal of any collision risk.
2. Rules 7(a), (b), and (c) state that every vessel must use all available means to determine whether any collision risk exists. Radar equipment, including long-range scanning, must be used properly to obtain early warnings of any collision risk and to undertake systematic observation of detected objects. Assumptions must not be made based on insufficient information.
3. Rule 7(d) states that a collision risk exists if the compass bearing of an approaching vessel does not change appreciably.
4. Rules 8(a) and (b) state that any collision avoidance action must be taken in accordance with this rule and in ample time with good seamanship. Any alteration of course and/or speed to avoid collision must be large enough to be apparent to the other vessel.
5. Rule 8(d) states that the collision avoidance action must result in passing at a safe distance. The effectiveness of the action must be carefully monitored until the other vessel is past and clear.
6. The first paragraph of Rule 19(d) states that in the case of restricted visibility, a vessel that detects the presence of another vessel by radar alone must determine whether a close-quarters situation is developing and/or any collision risk exists.

7. Rule 19(d) (i) and (ii) state that when a vessel detects the presence of another vessel by radar alone, it must not alter its course to port if the other vessel is forward of the beam except to overtake the vessel; moreover, it must not alter its course toward a vessel abeam or abaft.
8. Rule 19(e) also states that every vessel must reduce its speed to the minimum at which it can maintain its course when it receives the fog signal of another vessel forward of its beam or cannot avoid a close-quarters situation with another vessel forward of its beam.

Some academic research institutions, including universities, are required by the chemical hygiene plan (CHP) to develop and implement general standard operating procedures (SOPs) and specific SOPs for the particular chemicals handled by their laboratories. Safety in research at universities, etc. can be divided into two categories based on the time period during which people are affected: physical hazards, such as chemical explosions and fires, i.e. events that cause health hazards in a short period of time, and health hazards resulting from long-term exposure to chemicals, i.e. events that cause health hazards over a long period of time.

The former includes casualties caused by runaway reactions of energetic substances, fires and explosions caused by highly reactive chemical substances, retinal damage caused by laser light hitting the eye, gas poisoning caused by the diffusion of hazardous substances in confined spaces, acute health problems caused by the use of hazardous chemicals in quantities far in excess of regulated levels, and casualties and disappearances while diving at sea. The latter are cases where harmful substances, such as carcinogens and reproductive toxins, or noise from laboratory equipment, cause damage to human health over a long period of time.

For health hazards, minimum exposure limits are set by national occupational health and safety regulations. However, there are no target substances, research activities, or SOPs for dealing with accidents if they occur with regard to physical hazards. The author believes that the development of SOPs in collaboration with universities, etc. around the world will help to improve the situation where similar accidents occur around the world.

## 4.4 The Core of the Safety Management System

The SMS adopts the concept of the quality management system (QMS) and replaces the quality of the product and service under the control of the QMS with safety. Understanding the core concepts of the QMS is identical to operating the SMS correctly and effectively.

At the heart of the QMS is the Plan–Do–Check–Act (PDCA) cycle (see Figure 1.4 in Chapter 1), which must be repeated for continuous improvement. Iizuka (2009) describes an effective PDCA cycle that an organization should establish and operate as follows. It must be a process management approach that controls quality in each process of the PDCA cycle, rather than a first aid approach to individual defective products. Translated into safety, this means establishing a set of processes that ensure safety from the outset and managing each process to ensure overall safety, rather than a first aid approach to accidents at universities, etc. on a case-by-case basis. The following is an example of how the author has modified the SMS used in industry to make it suitable for universities, etc.

In addition, emergencies that should be dealt with at the organizational level include events such as fires, explosions, accidents, large-scale leaks or spills of hazardous substances, major power failures, major flooding, bomb threats, earthquakes, terrorist attacks, and pandemics. Organizations should prepare for emergencies by establishing business continuity management (BCM) and conducting drills during normal times. Contingency plans for these events implemented in the laboratory need to be integrated into the organizational response approach, or BCM, and there needs to be unity of intent so that all laboratories are aligned and have a consistent response approach.

For more information on BCM as a management strategy or system for universities, etc., the specific relationship between BCM and SMS, and the approach to embedding and operationalizing the system in the organization, see Chapter 8.

### 4.4.1 Plan

1. Clarify the objectives to be managed.

   This is the process of stating in writing what the organization's objectives are: for example, to ensure safety and environmental protection, to reduce the number of accidents in research and educational activities from the previous year, and not to cause health hazards due to exposure to hazardous substances.

2. Determine a scale to measure whether the objective has been achieved.

   If in step 1 above, the number of accidents is defined as the objective to be managed, the measure is the number of accidents that occur. Accidents can be categorized as fatal accidents, accidents resulting in serious injuries requiring hospital treatment for more than one month, accidents resulting in minor injuries requiring hospital treatment for less than one month, near misses that could have resulted in a serious accident in the nick of time, and incidents that did not result in death or injury but caused damage to equipment or other property. Near misses are events that could have resulted in a serious accident, such as fire, explosion, death or injury, gas poisoning, or lack of oxygen.

3. Determine the levels and targets to be achieved for the control subjects.

   In the case of accidents, for example, this means setting specific targets, such as reducing the number of fatal, major, and minor accidents by 10% each year and to zero in 10 years. In manufacturing, the Pareto chart is used to select control subjects. In quality control, 80% of product quality problems are attributed to 20% of causes. The idea is to make the 20% that is the cause of that product defect a control subject. This means that the organization has to select subjects that are statistically important.

4. Identify ways and means to achieve these objectives.

   Identifying ways and means is about providing safe work practices, procedures, and a safe working environment in education and research activities, including, but not limited to, the following:

   a. Comply with mandatory legislation and the bylaws of the local authorities in which universities, etc. are located.

      The area of research activities at universities, etc. is extensive and ranges from national legislation to local government bylaws. This includes workplace safety regulations, such as those of national Environmental, Health and Safety (EHS) authorities. In the United States, for example, the OSHA is the agency responsible for workplace safety and the regulations it issues. The regulation of certain chemicals is the responsibility of the fire marshal, so the fire code must be followed. Drugs and other substances used in animal experiments are regulated by the Drug Control Act. In some countries, diving for research purposes requires a license issued and controlled by the government. The university's EHS and compliance office must inform laboratories subject to legislation and monitor compliance with legislation and other regulations.

   b. Adopt applicable guidelines and standards recommended by organizations, governmental agencies, societies, and academic institutions.

For laboratory safety, a widely adopted guideline in the United States, *Prudent Practices in the Laboratory: Handling and Management of Chemical Hazards*, published by the National Academy of Sciences (2011), is useful.

**c.** Develop written procedures.

- Develop general SOPs applicable to field research activities and laboratory research activities.
- Develop specific SOPs for specific experiments.

  Universities, etc. handle a wide variety of chemicals and other unknown substances. Universities formulate a CHP to prevent accidents involving chemical substances, which should include specific SOPs in addition to general SOPs. When developing specific SOPs, it is advisable to incorporate the advice of experts or researchers conducting the same studies or experiments into the procedures.

**d.** Assess all identified risks related to people, equipment, and the environment and establish safety assurances. Including preparation for emergencies continuously improves skills related to the safety management of people.

- The following is an example of a chemical risk assessment. When scaling up highly reactive chemicals in experiments, the specific hazards of the chemicals handled should be investigated using the SDS, etc. For example, limit the amount of chemicals used to 50 mg or less, and as further risk reduction measures, use engineering controls such as fume hoods and glove boxes and equip with blast shields. Wear appropriate personal protective equipment (PPE) to minimize harm in the event of an explosion. Procedures should clearly state that emergencies can be dealt with, for example, by preparing an inhibitor to be used in the event of a runaway reaction.

  In the event of a fire, who is responsible for the initial firefighting depends on the fire regulations or organization in each country. In this section, the person who discovered the fire or a colleague will be responsible for the initial firefighting. Depending on the size of the laboratory staff, the laboratory may allocate a person to extinguish the initial fire with a fire extinguisher, a person to sound the fire alarm, and a person to direct colleagues and others to the evacuation exits. An emergency response program should be established, including the use of fire extinguishers appropriate to the type of chemical, and educational training to be given when there is a change in personnel due to a transfer. This program should include the location and use of safety equipment such as eyewash units and safety showers.

  In the case of toxic gases or highly toxic materials, the procedure differs from the above and includes immediate evacuation of the laboratory.
- Develop a program to ensure that equipment is inspected at statutory or appropriate intervals, such as diving equipment used in field research and laboratory equipment. Cases have been reported of researchers making their own equipment in the laboratory and this equipment causing accidents such as fires. A program is required to ensure that safety standards for equipment and methods to reduce risk to an acceptable level in accordance with the hierarchy of control are adhered to.
- Provide appropriate risk assessment training to janitors who collect waste from laboratory bins and external contractors who visit for laboratory repairs and maintenance. In addition, for visitors from other universities, minors, including children and members of the public visiting on tours of university laboratories, carry out a risk assessment of the area to be visited in advance, establish no-go zones, and determine safe tour routes.

**e.** Establish a system whereby accidents, incidents, near misses, and unsafe situations that do not conform to the procedures during an experiment are investigated and reported to the laboratory and the organization. Procedures that do not conform to the current situation must be revised and must be practicable at all times; these issues are addressed through the establishment and proper operation of the SMS.

**f.** The health and safety committee, in conjunction with the EHS, will develop an accident investigation analysis methodology and report writing procedures.

### 4.4.2 Do

1. The organization shall establish a program to identify the education and training required to support the SMS and shall ensure that all relevant personnel receive the necessary education and training. Relevant personnel include faculty members, researchers, and staff.
2. Provide education and training to researchers and students on procedures for the conduct of experiments.

   The following applies to the procedures to be followed for experiments with scaled-up quantities of chemicals as described in Plan (4) above. For example, trainers should provide education and training to trainees on how to identify chemical-specific hazards and why each process listed in the procedures is necessary, using each process of risk assessment and the laws of risk reduction.

   This is followed by practical, scenario-based training in the laboratory. Real equipment and chemicals should be used, and the process should be demonstrated step by step. If data on accidents or near misses are available, it is useful to explain in detail which process caused the accident or near miss, together with the causes of the accident, to help prevent recurrence.
3. Conduct experiments and other activities in accordance with procedures.
   - **a.** The researcher or student should seek the opinion of supervisors, colleagues, other researchers, and experts if there is any uncertainty in the procedures and should not proceed with an experiment in a situation where the procedures are unclear. Any ambiguities that have been resolved must be reflected in the revised procedures, made clear, and followed.
   - **b.** If the expected results are not obtained after following the procedure, or if a near miss occurs, the researcher must inform their laboratory supervisor and the EHS of the inadequacy of the operational procedure and the occurrence of the near miss. The EHS must review the reported deficiencies and correct the operational procedures in cooperation with the laboratory concerned, the health and safety committee, or the relevant technical committee. The changes shall be communicated to the laboratories concerned. This will prevent discrepancies between the general SOP or specific SOP and the work performed in the laboratory. For the handling of near misses, see Check process in Section 4.4.3.

### 4.4.3 Check

1. Inspect and check processes and results, based on factual information, to determine whether targets are being met and to examine whether there are any problems. Internal and external audits, management reviews, accident investigations, and laboratory walkthroughs fall under this process. Here, based on factual information means an evidence-based, scientific approach.

   Management review refers to the involvement of top management in the monthly workplace inspections of laboratories carried out by the health and safety committee. Top management includes the Provost, Dean of Research, etc. Top management will personally advise the laboratory faculty if they identify any EHS issues during workplace inspections, thereby demonstrating top management's commitment to EHS through their actions on the ground. The frequency of management reviews should be at least once a year, and management reviews are particularly important after an accident has occurred and to check that the safety recommendations published following an accident investigation have been implemented in the laboratory concerned.

2. Conduct regular workplace inspections for noncompliance with various regulations, university bylaws, general and specific SOPs, guidelines, and other procedures.
   a. The workplace inspection team should include members of the health and safety committee and faculty or experts in the relevant disciplines. Members of the health and safety committee may include Provosts, Deans of Research, Faculty, PIs, EHS, Facilities Department Engineers, Lasers, and other experts in specific areas. The results of workplace inspections are presented at monthly health and safety committee meetings. The health and safety committee must follow up on any issues raised during workplace inspections until they are resolved. The involvement of faculty members and experts as temporary team members in workplace inspection teams provides an opportunity for inspected laboratory staff to receive advice on unrecognized hazards, risk assessments, etc. In addition, to increase the effectiveness of workplace inspections, it is desirable that faculty, staff, and students associated with the laboratory being inspected be present during the inspection.
   b. Workplace inspections of research activities that take place off-campus, such as fieldwork, are carried out by the Fieldwork Research Safety Committee or the Workplace Inspection Team. The faculty member or PI in charge of the field experiment must demonstrate in the experiment plan, dive plan or other written plan they have prepared that the anticipated level of risk has been reduced to a level that the organization can tolerate and that the research team has prepared and is ready to implement a contingency plan by the date the experiment is to be conducted. One means of demonstrating this is for the PI to present an overview of the experiment and the safety-related implementation plan, dive plan, and emergency response plan to the Fieldwork Research Safety Committee or the Workplace Inspection Team and to obtain safety approval from the committee. The presentation includes discussion and advice on all aspects of the safety-related issues. After going through the process of this presentation, the points that need to be revised in terms of safety are incorporated into the plan, and then, the field research activities are approved in an organized way.
3. In order to verify that field research activities at sea and on shore are conducted in accordance with the implementation plan and procedures, that there are no deviations from the implementation plan and procedures, and that no problems have arisen, a diving safety officer shall randomly accompany the research activities to the site so that a third party can monitor and verify the research activities.
4. External audits of EHS shall be carried out by external experts on a regular basis. Independent of the external audit, an internal audit is carried out once a year by an audit team made up of employees of the organization. The results of the audit shall be presented to the faculty assembly, the risk management committee, the health and safety committee, and other committees attended by top management.
5. In the event of a serious accident, an accident task force headed by the President shall be set up to deal with the situation in an organized manner, and a third-party accident investigation committee shall be set up to conduct a comprehensive and scientific accident investigation and analysis. The third-party accident investigation committee comprises experts in accident investigation and experts in scientific fields relevant to the accident factors. The committee will conduct a comprehensive, evidence-based accident investigation to identify and analyze

accident-contributing factors using the scientific methods based on the approach outlined in Chapter 6. The EHS acts as the secretariat of the committee and supports the accident investigations.

The full version of the accident investigation report prepared by the committee shall be submitted to the accident task force, and the factors contributing to the accident and measures to prevent recurrence identified as a result of the analysis shall be communicated by the accident task force to all laboratory and organization personnel. The EHS, together with the health and safety committee, will follow up until measures to prevent recurrence are fully implemented in the laboratory where the accident occurred.

6. In the event of a serious accident, a moratorium on experiments should be imposed on all laboratories conducting the same type of experiments within the organization. The moratorium shall be announced and implemented by the Vice President for Research or Provost. The end of the moratorium will be implemented when a thorough accident investigation and analysis have been carried out to identify all contributing factors to the accident and these contributing factors have been fully rectified at the laboratory where the accident occurred.
7. In the event of a near miss that may result in a serious accident (hereafter referred to as a "serious near miss"), the EHS shall conduct an accident investigation analysis using the approach described in Chapter 6 and prepare a full version accident investigation report. The aim of this accident investigation is to identify all factors contributing to the near miss, to prevent the accident from happening, and to carry out proactive accident prevention.
8. Accidents other than serious accidents, such as when researchers accidentally cut their finger on a sharp object or get stuck by a needle during animal experiments, or when they slip and sprain their ankle while walking on a slippery floor, are investigated and reported in a manner different from the comprehensive accident investigation, analysis, and report writing that is carried out for serious accidents and serious near misses, as described in Chapter 6. The investigation and report will be prepared in a different way.

   In other words, the laboratory where the accident occurred will carry out a simple accident investigation based on a checklist and prepare a simple version of the accident investigation report.

The classification of accidents and incidents relevant to accident investigation is illustrated in Figure 4.1. This classification method focuses on the prevention of recurrence of accidents involving faculty, staff, and students

### 4.4.4 Act

1. During the Check process, action is taken when nonconformities or deviations from targets are identified as a fact. There are three types of actions.
   a. First aid measures: This is a temporary measure to remedy identified nonconformities on the spot and is the same as using a fire extinguisher to put out a fire if one breaks out. If a chemical reaction is about to run out of control, first aid measures are applied to stop the runaway reaction by cooling it down and treating it with a runaway inhibitor.

      If a colleague's behavior is observed to be abnormal while diving at sea, first aid measures include emergency responses such as a buddy bringing the colleague to the surface, taking into account the time required for the colleague to stop before surfacing, administering oxygen inhalation on board, and requesting transport to a medical facility, depending on the colleague's state of consciousness.

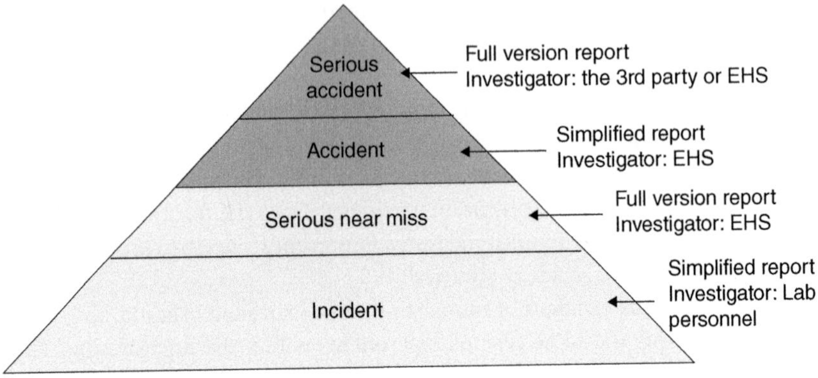


**Figure 4.1** Relationship between accidents and incidents as a classification for accident investigation. *Note:* The four main categories of event types are based on the emphasis on prevention of recurrence and proactive accident prevention. The relationship between the number of events is shown in the form of a pyramid based on Heinrich's law. To the right of the events are the type of report to be made and the person responsible for the investigation.

 **b.** Prevention of recurrence: Fire accidents will recur unless a thorough post-fire accident investigation is conducted to determine the cause of the fire. The method to prevent recurrence is to first identify all accident-contributing factors using a thorough and scientific accident investigation and analysis method to clarify why such a nonconforming event occurred. The identified accident-contributing factors shall be corrected in accordance with the hierarchy of control methods.
 **c.** Proactive accident prevention: Prevention of recurrence is a measure taken solely by the organization in which an accident, such as a fire, has occurred. There are many different types of accidents. In organizations where no specific type of accident has occurred, proactive accident prevention can be achieved by organizing the following actions: correction of unsafe conditions identified during workplace inspections and correction of factors contributing to the occurrence of near misses identified during near-miss investigation and analysis. Investigate and compare accident-contributing factors identified in accident investigation analyses at other universities and other academic institutions to see whether similar situations exist at your own university and correct these potential accident-contributing factors if applicable. Correction of issues identified, for example, by internal and external audits.

**2.** Ensure document management
 **a.** Everything implemented in the PDCA cycle, which is the core of the SMS, must be documented and stored.
 **b.** When documents are updated, expired documents must be destroyed, and a system must be put in place to prevent the mixing of old and new documents.

Emergency response plans developed by individual laboratories in accordance with organizational policy should identify a person responsible for a specific task, and a table should be posted with the name and role of the person responsible in the laboratory. In the event of an emergency, such as fire, explosion, injury or death, gas poisoning, lack of oxygen, or a large-scale release of toxic substances, people can become panicked and unable to act rationally. In 2018, a fire broke out at a building construction site in Tokyo when the flame from a gas burner used to cut steel caught fire in nearby polyurethane, filling the local workplace with smoke, killing five people left behind, and injuring 37 others (Nikkei Shinbun 2021). This was because the construction company had not clearly defined the procedures, roles, and evacuation routes to be followed by employees in the event of a fire. It is preferable for the emergency response plan to identify roles by the name of the individual rather than by the title of the job. When the person in charge or someone who plays an important role is transferred, it is necessary to decide on a new person in charge to simulate the procedure and ensure that the new person is familiar with the content of the procedure. By carrying out these document management and training activities, the roles and responsibilities of individuals become clear. In addition, training ensures that the person in charge is familiar with the procedures and is ready to respond whenever an emergency occurs.

## 4.5 Organizational Structure and Authority

### 4.5.1 General Description

Chapter 1 highlighted the structural and individual work differences between universities and industry. Given these differences, it is important to use methods that take advantage of the characteristics of the organizational structure of universities, etc. in order to construct an applicable SMS.

1. The purpose of the SMS shall be to prevent accidents and health hazards during education and research activities, to mitigate damage to research facilities and equipment caused by accidents, and to ensure protection of the environment.
2. The head of the organization shall make a commitment to the objectives.
   a. The head of the organization is the President or Chief Executive Officer.
   b. Top management must support safety recommendations, including the suspension of experiments, made by the Director of Safety Management and the Safety Management Branch, as well as other aspects such as required personnel, financial resources, and education and training programs. In this context, top management refers to the President and the Executive Directors. The National Academy of Sciences (2011) states the following about the commitment and responsibility of top management.

      *The ultimate responsibility for creating a safe environment and for encouraging a culture of safety rests with the head of the organization and its operating units... Even a well-conceived safety program will be treated casually by workers if it is neglected by top management.*
3. The organization shall clarify and document the responsibilities, authority, and interrelationships of all individuals or offices that manage, perform, and review tasks related to safety and environmental protection.
   a. The Director of Safety Management shall be an individual who can effectively implement improvements related to environmental health and safety in coordination with the faculty/PI and top management of the organization.
   b. The Director of Safety Management should be equivalent to the head of EHS in the general organizational structure of the university and shall be responsible for the programs established in the university-wide SMS. He/she has the responsibility and authority to ensure that each process of the PDCA cycle listed in Section 4.4 is implemented.
   c. The safety management of individual laboratories is the responsibility and authority of the faculty member representing that laboratory, Hereinafter referred to as the Laboratory Safety Manager. The Laboratory Safety Manager is responsible for laboratory safety and for

the safety of their staff and students. The National Academy of Sciences (2011) describes the role of faculty and PIs as follows:

*When each principal investigator offers leadership that demonstrates a deep concern for safety, fewer people get hurt. If any principal investigator projects an attitude that appears to be cavalier or hostile to the university safety program, that research group and others can mirror the poor example and exhibit behavior that sets the stage for potential accidents, loss of institutional property, and costly litigation.*

CSB (2010) also points out the following regarding organizational structure:

*The department Chair is considered the responsible person for ensuring safety; however, in practice, the Chair holds this managerial role while at the same time maintaining his/her role as a principal investigator for research; thus, a potential conflict exists due to the duality of the position. Authority and oversight of safety at a level above the Chair is a critical component of safety management within an academic structure.*

   **d.** The Director of Safety Management and the Laboratory Safety Manager shall provide the necessary education and training programs for environmental health and safety, including laboratory safety, to laboratory personnel and students. The laboratory personnel and students are required to participate in the education and training programs provided and ensure their safety during experiments and other activities.
   **e.** The Director of Safety Management shall explain matters relating to the SMS in education and research activities at top management meetings and at faculty assemblies.

**4.** Scope and applicability
   **a.** The policies and procedures laid down in the SMS shall apply to faculty, staff, and students of the organization concerned.
   **b.** Experiments may be conducted jointly with overseas teams. If the person responsible for the team is from an organization other than your university, your university's researchers will be under the safety control of the team and will be subject to its rules. The SMS at your university does not apply, but each researcher must take responsibility for his or her individual actions.

## 4.5.2 Establishment of Safety Management Regulations

In order to ensure that universities, etc. have the SMS in place and are able to operate it correctly, the objectives and operation of the SMS need to be fully discussed with the relevant departments and faculty members before proceeding. The administrative department in charge of safety management, usually EHS, needs to formulate a draft safety management regulation and obtain the President's approval. This section provides the following sample safety management regulations to facilitate and enable many universities, etc. to implement the SMS.

### 4.5.2.1 Sample of Safety Management Regulations

#### *4.5.2.1.1 Purpose*

***Article 1***

Once a casualty occurs in a laboratory or field experiment, not only is there a loss of life, but also a moratorium on all experiments or field work activities related to the accident, with enormous consequences for the continuation of education, research, and business at the university.

The purpose of these regulations is to establish a safety management system based on (fill in higher level regulations) to prevent the recurrence of accidents resulting in death or injury in the university's education and research activities and to prevent accidents before they occur.

#### 4.5.2.1.2 Definitions

***Article 2***

The definitions of the terms used in this clause are set out in the following paragraphs:

1. The Laboratory Safety Manager is the faculty member and section leader assigned to oversee the work associated with the experiment or field activity.
2. Affiliates and others are staff and students, or external participants involved in experiments or field activities in the laboratory under the direction and supervision of the Laboratory Safety Manager.
3. Safety considerations mean taking protective measures (safety measures) to prevent accidents from happening, including developing an emergency response plan and training for it, and taking measures to mitigate damage in the event of an accident.
4. Accidents are defined as events resulting in death and events resulting in serious injury requiring hospitalization or medical treatment for more than 30 days.
5. Incidents are events other than accidents, including near misses.
6. Unsafe conditions refer to inappropriate situations that may contribute to the occurrence of accidents. For example, it refers to situations where oxygen analyzers are not installed in laboratories where there may be a lack of oxygen, where chemical substances that must not be mixed are stored next to each other, and where regular inspections of diving equipment are not carried out.
7. Unsafe acts refer to inappropriate behavior that may contribute to the occurrence of an accident. For example, performing experiments with harmful chemicals without wearing appropriate PPE, diving in areas with fast tides and ocean currents where accidents have occurred in the past, and performing experiments without wearing safety glasses appropriate for the wavelength of the laser beam.
8. Contributing factor means a factor involved in the occurrence of a specific accident or incident.

#### 4.5.2.1.3 Scope of Application

***Article 3***

1. These regulations apply to laboratory experiments and domestic and international fieldwork activities in which the university is responsible for safety in education and research activities, which are primarily directed and implemented by the university.
2. These regulations apply not only to university faculty, staff, and students, but also to faculty, staff, and students other than the university who participate in educational and research activities for which the university has safety responsibilities.

#### 4.5.2.1.4 Safety Management System

***Article 4***

The laboratory safety managers, affiliates, and others have the following responsibilities when conducting experiments in the laboratory and educational and research activities in the field:

1. The laboratory safety managers shall prepare a plan of experiments or field research activities (hereinafter referred to as "the plan") that takes into account safety considerations in anticipation of foreseeable accidents. The plan shall be prepared in accordance with the following conditions.

2. The laboratory safety managers must ensure that affiliates and others receive the education, training, and medical examinations required for experiments or field research activities.
3. The laboratory safety managers must inform affiliates and others of safety issues and provide guidance and oversight during the conduct of experiments or field research activities to them.
4. Affiliates and others must use safe experimental procedures and safety measures. If in doubt about safety, consult the laboratory safety manager, or the affiliate, or other knowledgeable person.
5. Affiliates and others shall report to the laboratory safety manager when unsafe conditions or unsafe acts are detected in experiments or field research activities. If unsafe conditions or unsafe acts can be rectified on the spot, they must be reported after the rectification.
6. When the laboratory safety manager receives a report of unsafe conditions or unsafe acts from affiliates and others, he or she must correct them.
7. If an incident or accident occurs during an experiment or field research activity, the laboratory safety manager or affiliates and others must give priority to human life, take immediate action to mitigate harm, and then report the incident or accident to the EHS, which is the secretariat of the Safety and Health Committee. The incident must be reported in accordance with the Incident and Accident Reporting Guidelines administered by the EHS. In the event of a fatal accident, the accident must be dealt with in accordance with the Business Continuity Plan (BCP).
8. The laboratory safety manager must take action to correct contributing factors identified as a result of the incident or accident investigation carried out by the Safety and Health Committee.
9. The laboratory safety manager must incorporate the corrected contributing factors into the plan so that they are taken into account in the next experimental or field research activity to prevent recurrence of accidents.

***Article 5***

1. EHS must work with safety committees, such as the Safety and Health Committee, Fieldwork Research Safety Committee, Biosafety Committee, Radiation Safety Committee, and Laser Safety Committee (hereinafter referred to as "Committees"), to assess site safety and continuously improve safety by carrying out the following activities:
   i. In order to assess whether the safety-related issues described in the plan are being implemented in the field, random inspections shall be carried out.
   ii. If a site inspection reveals safety deficiencies, the laboratory safety manager shall be instructed to correct them if they can be corrected on the spot or, if it will take time to correct them, a deadline shall be set for correction and safety advice or recommendations shall be given to the laboratory safety manager.
2. The laboratory safety manager to whom the defect has been pointed out in Paragraph 1 (2) of this Article shall correct the defect by the designated date and report to the EHS.
3. The chairperson of the Safety and Health Committee that oversees safety must have each committee member report annually to the Safety and Health Committee on any deficiencies that have been rectified.
4. The EHS shall inform all people of the university of the contributing factors and lessons learned as a result of the accident investigation conducted according to Article 4, Paragraph 7 above, in order to prevent the same accident from occurring again and prevent any other accidents throughout the university.

#### *4.5.2.1.5. Detailed Provisions on Implementation*

***Article 6***

In addition to the provisions of these regulations, the Safety and Health Committee shall determine issues regarding operations of SMS.

##### 4.5.2.1.6 *Amendment and Abolishment of Rules*
***Article 7***
The President following a discussion by the Safety and Health Committee shall make amendment or abolishment of these regulations.

### 4.5.3 Establishment of Accident Investigation Guidelines

The accident investigation and analysis of accidents occurring at universities, etc. falls under the Check process of the safety management system. A wide variety of accidents and incidents occur, but it is physically difficult to investigate all accidents and incidents due to the limited manpower and budget of the departments in charge of accident investigations at universities, etc. Deciding which types of accidents should be thoroughly investigated and all factors contributing to the accident identified is an important decision in preventing the recurrence of accidents and in preventing accidents from occurring in the first place. The following guidelines have been developed to reflect the actual situation in universities, etc. For information on accident investigation and analysis methods, see Chapter 6.

#### 4.5.3.1 Sample of Accident Investigation and Reporting Guidelines

##### 4.5.3.1.1 *Purpose of These Guidelines*
In order to prevent the recurrence of accidents and to proactively prevent accidents from occurring in the first place, the health and safety committee (hereinafter referred to as "the Committee") will conduct accident investigations into serious near misses that could have resulted in fire, explosion, death, or injury at the university, as well as serious accidents that did result in fire, explosion, death, or injury. It then identifies the site conditions and accident-contributing factors that led to these events and communicates accident prevention measures to faculty members, laboratory staff, students, and relevant departmental staff. Accidents and incidents, other than serious accidents and serious near misses, will be recorded in a specified format by the person in charge of the laboratory where they occur and reported to the Committee, EHS, and the relevant department.

With the aim of preventing accidents in the first place, these guidelines provide comprehensive rules for reporting, evidence collection, analysis, report writing, and publication of incident and accident information within the university and other academic institutions.

##### 4.5.3.1.2 *Scope*
The accidents and incidents covered shall be the following that occur in relation to the university's educational and research activities.

1. Serious accidents: fire, explosion, fatal, and serious injury accidents. In this context, serious injury refers to injury that requires more than 30 days of treatment.
2. Accident: an event in which a person is injured.
3. Serious near miss: an incident which did not result in human injury, but which could have developed into a serious accident.
4. Incident: an event that did not result in human injury but caused damage to research equipment or other property.

##### 4.5.3.1.3 *Methods of Reporting*
Accident or incident information can be reported to the Committee using a web form. The Committee will follow up with the reporter to discuss corrective actions to prevent a recurrence.

Those who wish to make an anonymous report can fill in the designated report form and send it to the EHS using the internal mail system or in-house mail without their names on it.

In the event of an accident, the university emergency response center (telephone number xx) should be contacted. If first aid is required for an injured person, the health center (telephone number xx), or the National Emergency Services (telephone number xx) should be contacted first. However, within one month after the situation has calmed down, the injured person's supervisor should report the incident using the web form. Serious accidents and serious near misses will be investigated by an accident investigation team appointed by the committee. The accident investigation team consists of EHS staff.

#### 4.5.3.1.4 Accident and Incident Investigation

The person appointed by the Committee to be in charge of accident investigation within EHS (hereinafter referred to as "accident investigation personnel") shall do the following:

1. Upon receipt of an accident or incident report, confer with the reporter and visit the site to determine the circumstances under which the accident occurred.
2. Interview the reporter and the persons involved in the accident to determine the following:
   - Timeline leading up to the accident.
   - The circumstances at the time of the accident.
   - Emergency response following the accident.
   - Operating instructions or procedures for equipment related to the occurrence of an accident.
   - Safety measures, as determined by the reporter's laboratory, were in place at the time of the accident.
   - Other matters relating to the occurrence of the accident.
3. Photographs should be taken of the accident site and the equipment involved.
4. The risk management and quality management process approach model based on SHEL model and Swiss cheese model (see Table 5.1 in Chapter 5) should be used to collect evidence, and all accident-contributing factors must be investigated exhaustively.
5. If the event that has occurred constitutes an accident or incident, but the damage to persons or property is considered to be minor and the factors contributing to the accident are not likely to affect the university as a whole, the procedures set out in steps 1–4 above can be omitted, and only a summary of the accident can be kept as a record.

#### 4.5.3.1.5 Analysis of the Accident

The accident investigation personnel shall analyze the accident investigation based on the evidence collected as follows:

1. The chronological sequence of events leading up to the accident.
2. Measures taken after the accident, including emergency actions to mitigate damage and first aid.
3. The immediate cause of the accident (e.g. actions taken immediately prior to the accident in terms of the timeline).
4. Accident contributing factors that may be relevant to the occurrence of the accident (e.g. factors behind the occurrence of the accident).
5. Corrective measures to prevent accidents derived from steps (3) and (4) above.

   When formulating corrective measures, it is advisable to consult with the laboratory to which the reporter belongs and the relevant department when implementing the corrective measures and to combine emergency measures with fundamental measures.

#### 4.5.3.1.6 Qualifications of Accident Investigation Personnel

As the quality of an accident investigation report depends on the skills of the accident investigation personnel, the accident investigation personnel shall be trained in the following subjects:

1. Accident investigation and accident model.
2. SMSs and risk management.
3. Human factors.
4. On-site accident investigation techniques.
5. Interview techniques.
6. Accident analysis methods and formulation of corrective actions.
7. Accident investigation report writing.

#### 4.5.3.1.7 Content and Publication of Accident Investigation Reports

1. The accident investigation personnel shall prepare a report containing the following items:
   - A summary of the accident (if a chemical substance is involved, including the name and quantity of the substance, etc. not to exceed 200 words).
   - The circumstances leading up to the accident, the chronological sequence of events, and the first aid and other measures taken.
   - The immediate cause of the accident and any contributing factors which may be relevant to the occurrence of the accident.
   - Measures to prevent recurrence of accidents (emergency and fundamental measures).
   - Data and photographs relating to the accident.
2. The results of a thorough investigation of the Software, Hardware, Environment, Central Liveware, and Peripheral Liveware using the SHEL model shall be described.
3. The subjects of the accident investigation report should include a summary of the accident, factual information, analysis, conclusions, and safety recommendations.
4. The accident investigation report shall be evidence-based, the analysis shall be derived from the factual information, and the conclusions shall be drawn from the analysis. Safety recommendations shall be designed to rectify accident-contributing factors identified as a result of the analysis.
5. The contents of the report will be made available to faculty, staff, and students involved in the accident via the website, but information, such as unit names, personal names, and other personal information, as well as cutting-edge technical information, will not be made public. The reports will only be used for accident prevention activities and reporters will not be penalized for reporting accidents.
6. Some anonymized reports can be shared by inter-university organizations that, with the provost's agreement, agree to their use for the purpose of preventing accidents and ensuring safety during educational and research activities.

### 4.5.4 Establishment of Business Continuity Plan

In the event of a crisis, major disaster, serious accident, or act of terrorism, normal university operations will be suspended. In particular, if an accident occurs during an oceanographic experiment in the field, the activities of laboratories conducting similar oceanographic experiments may be suspended for more than a year until the contributing factors of the accident are clarified through a comprehensive accident investigation and corrective measures are taken for the same accident-contributing factors. The BCP is a systematically formulated plan to ensure the early resumption of education and research activities, while at the same time preventing the research team on-site from being caught up in a secondary disaster.

Serious accidents, such as the ones resulting in fatalities or serious injuries, that have a significant impact on the university's operations are addressed in the BCP. In this BCP, an accident investigation committee consisting of external experts is established to conduct a thorough accident investigation and to identify all factors contributing to the accident, including organizational factors. The accident investigation team established within the university and led by the EHS is not suitable for investigating organizational factors due to conflicts of interest and will be carried out by a third-party committee outside the EHS. In other words, accident investigations must objectively examine and analyze organizational factors such as the formulation of general SOPs and the status of training and education, the conduct of workplace inspections and the follow-up status of the identified safety issues, and the status of the accident or near-miss investigation system. This is because the EHS has the responsibility and authority to carry out these tasks and therefore becomes the subject of the investigation itself.

#### 4.5.4.1 Sample of the BCP for Accidents Involving Death or Serious Injury During Education and Research Activities

The BCP will be activated in the event of a fatality or serious injury requiring more than one month of treatment during laboratory or field research activities and it aims to minimize the human impact on the research team, prevent secondary disasters, implement measures to prevent recurrence, and resume education and research activities as soon as possible.

When an accident involving a death or serious injury occurs, it is important to notify emergency services, such as the fire brigade, police, and coast guard, take immediate measures to mitigate the damage, provide support to the families of the victims, conduct an accident investigation, and implement corrective measures to prevent a recurrence of the accident.

##### *4.5.4.1.1 Basic Policy*

In the event of an accident, efforts shall be made to minimize damage through first aid measures and the prevention of secondary accident. In addition, factors contributing to the accident should be investigated and measures to prevent recurrence should be formulated as soon as possible so that a safe education and research environment can be quickly established.

##### *4.5.4.1.2 Objectives*

1. To ensure the safety of faculty, staff, and students during their educational and research activities and to minimize damage in the event of accidents.
2. To limit the interruption of educational and research activities to the shortest possible time by carrying out early investigation and analysis of the factors contributing to the accident and by taking effective measures to prevent recurrence.
3. To provide prompt assistance to accident victims and their families and to respond to their requests.

##### *4.5.4.1.3 Establishment and Organizational Structure of the Task Force*

In the event of a fatal accident or an accident involving one or more seriously injured members of the university's faculty, staff, and students, the Task Force shall be established.

The head of the Task Force shall establish an Accident Investigation Committee consisting of external experts, taking into account the impact of the accident on the university's operations. The Accident Investigation Committee shall investigate and analyze the accident and submit an accident investigation report to the head of the Task Force in order to prevent the recurrence of similar accidents and to strengthen the university's safety management system. Each team under the command of the Task Force designated in the organizational structure shall be responsible for the preparation and retention of records relating to the implementation of the response to accidents resulting in death or serious injury.

#### 4.5.4.1.4 Phase 1: Action in the Event of an Accident Involving Casualties

In the event of an accident, the research team involved shall report the incident to the emergency services and the university via the prescribed emergency communication network.

1. Faculty, staff, and students who are involved in an accident during their teaching or research activities should immediately call the fire brigade and the police on land and the coast guard at sea to report the accident.
2. The caller or accompanying person shall follow the instructions of the rescue services in order to minimize damage by providing first aid or to prevent secondary disasters, such as an accident involving a colleague on the way to the rescue.
3. After calling the emergency services, the caller reports the situation at the scene and a summary of the accident to the emergency contact person (the responsible department at the university).
4. The emergency contact person reports the situation at the scene and a summary of the accident to the General Affairs Team and the relevant departments (e.g. Academic Affairs, Human Resources, the department to which the casualty belongs) via the prescribed emergency contact network.
5. If the casualty is informed of the accident by the emergency services or through media reports, start the next process, "Assessing the accident and setting up the Task Force."

#### 4.5.4.1.5 Phase 2: Assessing the Accident and Setting Up the Task Force

Set up the Task Force and send the General Affairs Team and others to the site.

1. The General Affairs Team shall inform the members of the Task Force and the heads of each group that the Task Force has been set up and that they should assemble in the designated meeting room.
2. After receiving a report on the accident from the General Affairs Team, the head of the Task Force shall set up the Accident Investigation Committee consisting of external experts and dispatch staff from the General Affairs Team and relevant departments to the accident site as necessary.
3. On arrival at the accident site, the dispatched staff shall report successively to the Task Force on the situation at the site, including the actions of the rescue and investigation services.
4. The First Aid and Health Team shall provide psychological assistance to those involved in the education and research activity in which the accident occurred.
5. The department responsible for the implementation and management of the education and research activity concerned (hereinafter referred to as the "implementing department") shall provide detailed information to the relatives of the victim.
6. The General Affairs Team, in cooperation with the implementing department and relevant departments, shall provide support measures in response to requests from the family members of the victim.
7. The Public Relations Team, in cooperation with the Task Force and the General Affairs Team, shall respond to media interviews.

#### *4.5.4.1.6 Phase 3: The Accident Investigation Committee and Preventing Recurrence*

In addition to supporting the families of the victims, the Task Force receives the accident investigation report submitted by the Accident Investigation Committee and decides on improvement measures to be taken to prevent recurrence and strengthen the safety management system.

1. The Accident Investigation Committee shall visit the scene of the accident, cooperate with the rescue and investigating authorities, and investigate the detailed circumstances and contributing factors of the accident.
2. The Accident Investigation Committee shall, as appropriate, report to the General Affairs Team the information obtained during the on-site investigation. If, in the course of the accident investigation, the Accident Investigation Committee identifies matters requiring safety recommendations to prevent the occurrence of similar accidents, it may make safety recommendations to the Task Force through the General Affairs Team without waiting for the submission of the accident investigation report.
3. The Accident Investigation Committee shall submit an accident investigation report to the Head of the Task Force within a target period of three months after the accident, containing a summary of the accident, factual information, analysis of the accident, factors contributing to the accident, and safety recommendations.
4. Based on the results of the accident investigation, the Task Force shall decide on the improvement measures to be taken to prevent recurrence and strengthen the safety management system. The corrective actions shall specify the departments subject to these actions, the matters to be improved, and the deadline for implementation of the improvements.
5. The Task Force shall be disbanded when the process described in step 4 has been completed.
6. The General Affairs Team will continue to provide support to the families of the victims in cooperation with the implementing unit and the relevant departments.
7. The General Affairs Team, in cooperation with the relevant departments and committees established at the university after the accident, will disseminate the improvement actions and follow up to ensure that the improvement actions are implemented and will report on the current situation at regularly convened health and safety committee meetings.

## 4.6 Conclusion

Reanalysis of accident investigations or published accident investigation reports of accidents that occurred in university chemistry laboratories or during field research activities at sea or in rivers have shown that risk assessments, such as hazard identification and risk analysis, were not properly carried out and that SMSs in laboratories and universities were inadequate.

This chapter looks at the risk management processes and the SMS needed to prevent accidents and mitigate damage at the organizational level in universities, etc. In universities and academic institutions, risk assessment as laboratory safety is a subject that should be implemented as part of education and training and the need for it is recognized, but the risk management process and safety management system are not familiar concepts and the need for them is not fully recognized.

In the risk management process, hazard identification, risk analysis, risk evaluation, risk treatment, monitoring, and review have been described in detail and adapted to the actual situation in universities, etc., as they are carried out in other industrial sectors. In many universities and research institutions, risk assessments are already well-established and carried out in the laboratory. Nevertheless, accidents have occurred in laboratories as a result of inadequate risk assessment. In the area of human error, since it has been scientifically proven that an error made by people cannot be detected by themselves, the author proposed a method to avoid single-person experiments and to perform them with more than one person so that errors do not directly lead to accidents. For example, he described a thorough implementation of the risk management processes used in other fields, as demonstrated in Chapter 2, in chemistry laboratories, and in field research activities. This means that the researcher carries out the experiment with more than one person, and a colleague, supervisor, or PI who is familiar with the experiment being carried out is present at the experiment to monitor the process. The approach is that if the researcher deviates from the protocol or SOPs and takes inappropriate action, the person present will give advice to correct the course of the experiment and steer it in the right direction.

The maritime industry contributes to the prevention of collision accidents by having practices for the prevention of marine casualties at sea developed as international rules and implemented on the oceans worldwide. The procedures implemented in this other industry are general SOPs that incorporate risk management processes into the operation of ship collision avoidance and are considered to be helpful in improving safety in laboratory and field research activities conducted by universities, etc.

As a method of accident prevention, if the factors contributing to accidents are identified through scientific and comprehensive accident investigation and analysis at other universities, and if it is found that the same factors exist at the reader's university, etc., immediate action should be taken. In other words, it is recommended that potential accident-contributing factors existing in the reader's universities, etc. are corrected in accordance with the risk management process, including the hierarchy of control described in Chapter 2.

In the SMS, CSB (2010), OSHA (2013), and others in the United States, following an accident involving an explosion during a chemical experiment at University A in 2010, expressed the need to implement an SMS adapted to the actual conditions of the university rather than to implement the SMS used in industry as it is in the university.

This chapter describes a specific SMS that can be applied to universities, etc., reflecting the organizational structure and actual conditions of the university. One of the background theories of the Swiss cheese model discussed in Chapter 1 leads to the theory of building multiple layers of defense in an organization to prevent a single error from escalating into a major accident. The importance of implementing the SMS is that, in addition to the layers of defense of the risk assessment and risk management process, further layers of defense exist and operate as a system within the organization. As evidenced by real examples from other industries, notably the aircrafts collision at the Haneda Airport in January 2024, in which not a single passenger or crew member, including infants, was killed, it is believed that this SMS will prevent serious accidents where a single operator error does not lead directly to human casualties and the accident trajectory is blocked by multiple layers of defense.

## References

Fukuoka, K. (2019). *Safer Seas: Systematic Accident Prevention.* Boca Raton: CRC Press Taylor & Francis Group.

Iizuka, Y. (2009). *Introduction of Modern Quality Management.* Tokyo: Asakura Publishing Co. Ltd. (Japanese).

International Maritime Organization (2000). Guidelines for voyage planning. *Resolution A* 893 (21).
International Maritime Organization (2011). Revised Recommendations for Entering Enclosed Spaces Aboard Ships. *Resolution A* 1050 (27).
IEC/ISO 31010:2009 (2009). *Risk Management-Risk Assessment Techniques*. Geneva: International Electrotechnical Commission.
ISO 31000:2009 (2009). *Risk Management-Principles and Guidelines*. Geneva: International Organization for Standardization.
ISO/IEC Guide 51: 1999(E) (1999). *Safety Aspects-Guidelines for their inclusion in standards*. Geneva: International Organization of Standardization and International Electrotechnical Commission.
Maguire, R. (2006). *Safety Cases and Safety Reports: Meaning, Motivation and Management*. Surrey: Ashgate Publishing Limited.
National Academy of Sciences (2011). *Prudent Practices in the Laboratory: Handling and Management of Chemical Hazards*. Washington, DC: The National Academies Press.
Nikkei Shinbun (2021). Two construction workers have been found guilty in the Tokyo building fire that killed five people. *Nikkei Shinbun* (16 December).
Occupational Safety and Health Administration, USA (2013). Appendix A to §1910.1450-National Research Council Recommendations Concerning Chemical Hygiene in Laboratories (Non-Mandatory).
Okinawa Prefectural Government (2024). Beware of hubs. Okinawa Prefectural Health and Medical Care Department, Health and Drugs Division. https://www.pref.okinawa.jp/kurashikankyo/petgaiju/1018721/1005847.html (accessed 2 May 2024).
US Chemical Safety Board (2010). Texas Tech University Laboratory Explosion, No. 2010-05-I-TX. https://www.csb.gov/texas-tech-university-chemistry-lab-explosion/ (accessed 10 May 2023).

# 5

# Human Factors

## 5.1 Introduction

This chapter discusses human factors, which are necessary for understanding the factors behind accidents that occur in universities, etc. Human factors were first used in aircraft accident investigation analysis, and its definition has been subject to various interpretations and definitional difficulties because it encompasses many disciplines, including psychology, engineering, physiology, medicine, sociology, and anthropometry. The human factors are not only factors related to people but also include social components covering many disciplines. The ICAO has included the software, hardware, environment, and liveware (SHEL) model, as a graphical representation of the relationship between people and their environment, in its Accident Investigation and Analysis Manual as a model that can explain human factors.

In this section, the human factors that need to be learned in order to carry out accident investigation and analysis in universities, etc. are selected from some of the past accidents and incidents that actually occurred in universities, etc. and are explained for each element of the SHEL model. This will enable you to find out exactly what thc human factors are. As the scientific investigation and analysis of accidents in universities, etc. is not yet very advanced; the findings from other fields will be modified and explained so that they can be applied to the field of accidents in universities, etc.

This chapter, when read in conjunction with Chapter 7, "Contributing Factors Found in Accidents," will provide the reader with specific topics for in-depth accident investigations for Environment, Health and Safety (EHS) conducting investigations when accidents occur in universities, etc. It will also provide new perspectives of insight into accident and safety efforts for faculty, researchers, and students in laboratories and field research activities working on accident prevention.

## 5.2 Background

The concept of human factors began during the Second World War, when the design of military aircraft in Europe and the United States took into account human characteristics and limitations, which enhanced the maneuverability and combat performance of military aircraft and influenced the conduct of operations. This concept was adopted by ergonomics and applied to aircraft and ship design. It was only in the second half of the twentieth century, as discussed in Chapter 1 (see Section 1.8), that people began to realize that human factors could be used to prevent accidents,

*Accident Prevention and Investigation: A Systematic Guide for Professionals, Educators, Researchers, and Students*, First Edition. Koji Fukuoka.

and as socio-technical systems developed and industrial structures became more complex, the causes of accidents that occurred in them became more complex. Accident investigation and analysis, which until then had focused on human error, was no longer able to see the whole picture of accidents and identify the faulty elements, so there was a need for a way of thinking that could encompass the social components (Fukuoka 2019).

The SHEL model was developed in the field of aircraft, and prior to its introduction, it was believed that the human characteristics of individual pilots or crew members had a significant impact on the performance of aircraft operations. The SHEL model expands the scope of accident factors to include the teamwork of the crew, including the cabin attendants, and the management of the airline and air traffic control staff. It places the human element at the center of the model, showing conceptually and graphically how accidents occur through the interaction of other elements placed around it, namely, SHEL and peripheral liveware (Lp). The SHEL model has also been adopted by national accident investigation authorities in the field of marine casualties, as it provides an approach to the systematic collection and analysis of accident-related evidence.

## 5.3 Aims and Scope of Human Factors Investigations

ICAO (2001) published the *Aircraft Accident and Incident Investigation Manual*, which advocates a systematic approach based on the concept of human factors that is both efficient and effective and includes specific content.

According to the manual, the objectives of human factors investigation and analysis are to improve the safety of aircraft operations by:

- Identifying how human behavior contributed to the cause or development of an accident.
- Identifying the safety implications of limitations in human capabilities and behavior using physiology, psychology, sociology, ergonomics, etc.
- Making safety recommendations to prevent a recurrence of the same accident or to mitigate damage from accidents resulting from incorrect actions or decisions by individuals or groups involved in the accident.

In order to achieve the above objective, the manual states that human factors investigation and analysis should extend beyond the aircraft crew to include investigation and analysis of individuals or organizations involved in the accident, such as management, manufacturers, and government regulators.

Given that the purpose of accident investigation is to prevent recurrence and mitigate damage, the above purpose is likely to be the same as that of accident investigation and analysis in other sectors that wish to reduce the number of accidents that occur, and it seems inevitable that the same model will be adopted in other sectors. Here, safety recommendations refer to safety issues that are not legally required but are strongly recommended.

## 5.4 SHEL Model Applicable to Accidents at Universities, etc.

Hawkins (1987) published a modified version of the SHEL model, which was developed by Edwards in 1972 and revised in 1984. At that time, the concept of a safety management system (SMS) and its importance in relation to accident prevention had not been fully recognized. Therefore, this chapter supplements the Lp of the model by adding the content of SMS and modifying the software to include the procedures, practices, and emergency responses actually in use at the local workplace at the time of the accident. This clarifies how the organization's safety management system was operated in the local workplace and how the discrepancy between the SMS and the actual situation contributed to the accident.

Furthermore, it enables a detailed investigation and analysis of the situation of interrelationship between human factors and surrounding people and objects, including software, hardware, environment, attributed organizations, relevant administrative and private institutions, etc., related to the accidents, so that the problems of interrelationship can be identified (Fukuoka 2019).

Table 5.1 at the end of this section corresponds to Table 2.1 of the risk management and quality management process approach model, which consists of the chart shown in Figure 3.5 in Chapter 3, and allows for the comprehensive collection and analysis of accident-contributing factors occurring at universities, etc., using the SHEL model. Table 5.1 supplements the table published in 2022 (Fukuoka and Furusho 2022) with higher level accident-contributing factors, which show causal relationships between accident-contributing factors occurring at local workplaces. These are accident-contributing factors for which safety recommendations are made to take action to prevent recurrence. For each topic in Table 5.1, safety recommendations must be formulated more specifically to suit the actual circumstances of each accident.

The elements of the SHEL model, including the modified topics, are explained below.

### 5.4.1 Software/Procedures

Software describes the relationship between liveware and software and reflects the relationship between the people in the local workplace and the support systems or nonphysical parts of the system present in the workplace. It includes rules, protocols, standard operating procedures (SOPs), manuals, checklists, diagrams, maps, charts, publications, and computer software design. These documents include essential qualifications and certifications required for the task, as well as guidelines and standards issued by industry associations. The interrelationship between liveware–software includes whether standard operating procedures and other documents are available at the local workplace, whether these documents are stored in a way that allows users to access them, and whether the content of the documents is consistent with actual work processes.

Parrott (2011) states that when transitions occur, the environment becomes more prone to accidents. Transitions refer to periods when the environment or a person's shift in a rotating shift system changes from one state to another, reducing the operator's level of situational awareness. The following situations in which this transition occurs are considered in the context of accidents at universities, etc. (1) change of shift; (2) change in the workers, researchers, and leadership working together; (3) change of organization; (4) change of experiment/work plan; (5) change of equipment configuration (e.g. change of communication channel); and (6) change of experimental equipment (e.g. change to a different type of experimental equipment).

#### 5.4.1.1 Situation at the Scene of the Accident

##### *5.4.1.1.1 Local Workplace Procedures at the Time of the Accident*

If experiments are carried out, they should be conducted according to laboratory procedures and protocols at the local workplace, i.e. the laboratory or field research site. The committee conducts review for which the experimental plan requires prior approval by the relevant committee before experiments. For field research activities, the Fieldwork Safety Committee and other committees also review plans for field activities in advance, and the researcher is expected to revise any problems that are pointed out and conduct the experiment in the field with the approved experimental plan. Many accident investigations have shown that these procedures and protocols were not followed, or that the procedures and protocols were missing parts to ensure safety, or that the protocols themselves did not exist.

There have been many cases in other industries where inadequate procedure manuals and other documents have been a factor in accidents and statistics have been published. Reason (2008) found that about 70% of human performance problems in the nuclear industry were related to procedures. These procedure manuals contained incorrect information, inappropriate and unworkable situations, outdated information, and unlisted relevant tasks. In addition, some of these procedure manuals could not be accessed by workers because it was unclear where they were stored.

The results of Reason's research on the above procedure manuals can help to identify deficiencies or inadequacies in the manuals for researchers and accident investigation personnel. When using a procedure manual that is referenced in an authoritative book in the field or generally accepted by an academic or other organizations, some of the backgrounds of the accident can be identified by investigating the relationship between the process listed in the procedure manual and the accident. In other words, the procedures and protocols used in laboratories and fieldwork research activities are broken down into risk management processes, and it becomes clear which processes are prone to accidents and what the accident factors are.

Although the above is from the perspective of the person investigating the accident, the researcher conducting the experiment may recognize that deficiencies or inadequacies in the procedure manual for the experiment may lead to accidents. This allows researchers to determine which processes they should focus their attention on during accident prevention training.

In the maritime sector, there are international regulations that require navigators to constantly monitor the movements of other vessels approaching at sea during a voyage and to take collision avoidance measures in accordance with the Convention on the International Regulations for Preventing Collisions at Sea (COLREGs). Fukuoka (2015) analyzed 54 marine accidents in the risk management process and found that approximately 70% of the accidents, whether day or night, were caused by collisions as defined in the COLREGs that occur in the risk analysis process. It was also found that the reason for not carrying out the risk analysis required by the international rules was that the navigator's assumption that the other vessel would change course and avoid his vessel intervened in the decision-making process and facilitated to violate the international rules.

**5.4.1.1.1.1 Different Levels of Risk for Different Groups of Researchers** In terms of procedures, experiments with hazardous chemicals and other substances naturally involve different levels of risk and different risk controls for undergraduates, postgraduates, and faculty. The chemical substances handled by undergraduate students are already known in terms of their properties, and risk assessment when conducting experiments is documented as an effective and well-known method. Occupational Safety and Health Administration (OSHA 2013) and the National Academy of Sciences (2011) recommend that experiments with hazardous chemicals should be conducted in groups rather than alone. For example, supervisors or graduate students and trained laboratory personnel are expected to conduct experiments together. Therefore, if the experiment is carried out according to the protocol, safety is guaranteed. Accidents occur when some procedures in the protocol are omitted.

However, postgraduate students and faculty members are dealing exclusively with chemical substances whose properties are unknown or compounds produced from them in the experiment. These substances have no published precedents or previous research, so there are no procedures or documentation to follow during the experiment. Their experiment is in the same situation as a so-called trial and error experiment. There are two aspects to trial and error. An error made during an experiment may inadvertently lead to a discovery that contributes to a breakthrough in science. In addition, an error can turn into an accident.

Risk assessment consists of three processes: (1) hazard identification, (2) risk estimation, and (3) risk evaluation. In the case of undergraduate students, the frequency of accidents, or the frequency scale, is small if the experiment is carried out as described in the procedure. However, in the case of postgraduate students and faculty members, if it is assumed that the probability of accidents is the same as the probability of two sides of a coin, then accidents will occur 50% of the time in each experiment, which means that the frequency scale is large.

As a method of preventing accidents when handling chemicals of high or unknown risk, the National Academy of Sciences recommends experiments in units of 25–100 mg for solids and 100–200 μL for liquids. It also states that experiments with materials that can cause explosions during testing, such as energetic materials, should be accompanied by risk reduction and mitigation measures, including the use of engineering controls such as chemical fume food, blast shields, etc. (National Academy of Sciences 2011).

If there are alternative chemicals that can be substituted in experiments to achieve the same effect, using this approach means adopting the concept of inherently safe design in the hierarchy of control and the level of risk can be significantly reduced.

**5.4.1.1.1.2 Use of PPE** Performing a risk assessment that includes the use of personal protective equipment (PPE) can reduce the level of risk, but there will still be residual risks. It should be noted that PPE is a step taken as a last measure in the hierarchy of control, and the type of PPE to be worn will depend on the level of risk. The glove, goggles, mask, or laboratory coat to be worn will depend on the characteristics of the substance being handled in the experiment; the appropriate PPE should be clearly identified in writing by the laboratory faculty or manager, and PPE should be provided, for example, at the entrance to the laboratory to ensure that it is worn.

**5.4.1.1.1.3 Emergency Response Plans** An emergency response plan must be developed for any risk assessment carried out at a university, etc. The emergency response plan should cover not only the responses to chemical spills, explosions, and fires during experiments – whether in the laboratory or in the field – but also first aid procedures and emergency communication networks with the emergency or rescue services and the university. These measures are essential for mitigating injuries or illness to researchers during field research activities.

As a form of administrative control, oxygen concentration or gas detectors and alarms shall be installed to detect the escape of harmful substances into the laboratory and to inform the surrounding area. Consideration should also be given to the installation of fire extinguishers appropriate to the characteristics of the ignited material in the event of a fire and to the provision of two escape routes from the laboratory to the emergency exit of the building.

### 5.4.2 Hardware/Equipment

Hardware refers to the liveware–hardware interface that represents the interaction between man and machine. A prominent real-world example of the liveware–hardware interface in the military and aircraft sectors is the analog display of the three-pointer altimeter in the cockpit, which led to a series of accidents in which pilots misread the altitude and crashed. Replacing the altimeter with a digital display has eliminated similar accidents. In the marine sector, the International Maritime Organization (IMO 2000) established guidelines for man–machine interfaces for bridge arrangements and equipment, providing ergonomic designs for optimal positioning of displays, controls, etc. With regard to the placement of displays and controls, the most important or frequently used displays should be within 30° of the center of the user's preferred field of vision. The most important and frequently used controls must be located in the most convenient position for easy reach and grasp. Displays and controls must be adapted to the logical arrangement of people and grouped together in functional groups. Guidelines have been developed to ensure that the communication facilities, steering equipment, and machinery on a ship's bridge are designed and arranged based on ergonomics.

#### 5.4.2.1 Condition of Equipment, Tools, Etc. in Relation to the Accident

##### *5.4.2.1.1 Human–Machine Interface*

Depending on the purpose of the experiment, the laboratory is equipped with a wide variety of machinery, equipment, and instruments of different types and sizes, including fume hoods, safety cabinets, laser equipment, and high-pressure gas equipment. For machinery and equipment, the design of workstations, displays, controls, seating, and all physical parts of the system must be adapted to human characteristics. Some products use ergonomic design, such as placing the monitor at the researcher's eye level. In some cases, researchers use home-made equipment, but in such cases the design of the equipment must take into account ergonomic issues, such as designs that make it difficult to access the faulty part with tools or designs that are prone to installation errors.

When conducting fieldwork research in the marine environment, equipment such as regulators, buoyancy control devices, dive computers, masks, snorkels, fins, wet suits, and dry suits are the subject of hardware-related accidents. For wilderness and mountain expeditions, this includes GPS-equipped devices for accurate positioning, compasses, and communication equipment.

As technology has advanced, the design and materials of laboratory equipment have improved, with machines and digital technology performing tasks that were once performed by humans. The capacity of human sensory organs is enormous and can gather a lot of information, but the speed at which information is transmitted is limited by the process of understanding, predicting, and making decisions about the events under consideration, and there is only one channel and slow processing capacity.

According to Hawkins (1987), hardware must be designed for human characteristics, including physical, visual, auditory, tactile, and thought processes.

##### *5.4.2.1.2 Equipment and Instrumentation Status*

With regard to the equipment used in experiments, there have been a number of cases where accidents have occurred due to a lack of prior inspection of the equipment, such as cracks appearing in the equipment before the experiment. In order to prevent such accidents, it is necessary to carry out a risk assessment of the equipment and instruments during the preparation phase of the experiment.

##### *5.4.2.1.3 Alarms, Oxygen and Gas Concentration Detectors*

Alarms and detectors that warn people of impending harm are under administrative control in the hierarchy of control, including portable ones, and can act as risk reduction measures if used appropriately. In some chemical plants, fatal accidents occurred during operations where toxic gases were expected to be generated because warning devices were not installed. This was due to the gas detector malfunctioning due to silicon buildup on the contact points, so the device was removed. However, technological advancement has led to the development of new devices that do not allow silicon to adhere (US Chemical Safety Board [CSB] 2021). Researchers also need to keep up to date with the state of the art in warning devices that can reliably warn of a hazard.

### 5.4.3 Environment/Atmosphere

Environment refers to the liveware–environment interface, where education and research activities at universities, etc. take place not only in the laboratory but also outdoors in rivers, oceans, and mountains. The liveware–environment interface is therefore the relationship between people and the environment of the laboratory or experimental facilities and between people and the natural environment.

#### 5.4.3.1 The Atmosphere of a Laboratory

Accident-contributing factors relating to the laboratory environment include the following situations. Conducting experiments or preparations without being aware of oxygen concentrations below 18%, which can lead to oxygen deficiency or other symptoms; gas concentrations of certain substances, which can lead to gas poisoning; lower and upper flammable limits of certain gases, which can lead to explosions and fires.

In the event of a fire or explosion, damage may be exacerbated by failure to properly secure high-pressure gas storage tanks, leaving unnecessary flammable materials near experimental equipment, and poor laboratory organization. Power sources for electronic equipment and devices, including mobile phones, can be ignition sources and accident-contributing factors in fires and explosions.

Handling of carcinogenic and organ-damaging chemicals must comply with the statutory permissible exposure limits. Permissible limits for inadequate lighting, vibration, noise, etc., which cause long-term health problems, must also be observed.

It is important to ventilate the laboratory, which is an established method of engineering control. Ventilation methods include the use of chemical fume hoods and the installation of ventilation equipment.

#### 5.4.3.2 The Weather and Sea Conditions

In the ocean, the conditions of the marine environment, such as the speed of the tides and currents and the height of the waves, are also important factors. Other factors include the condition of the seabed topography, the clarity of the seawater, and the traffic of ships, jet skis, and other vessels in the experimental area.

There are several ways of conducting research at sea, such as boarding a vessel belonging to a research organization and sailing to the experimental area, or in some cases the researcher maneuvers a boat or hires a local vessel with a captain. Research and analysis of accident investigations in the maritime industry have shown that there are areas of the sea where ship's collision accidents are more likely to occur; attention should be paid to areas where marine traffic flows from three directions are aggregated and dispersed, especially in areas where the area is a waypoint for vessels (Fukuoka 2019).

The presence of marine organisms, wild animals, and plants that may be harmful to humans should be examined in advance to avoid approaching these organisms, animals, and plants and to prepare first aid in case of injury.

For marine experiments involving scuba diving, the proximity of medical facilities that can deal with decompression sickness or the bends and whether there are areas where mobile phones can be used are necessary issues that should be investigated in advance in order to implement damage mitigation measures.

In mountainous areas, the main environmental factors are the difference between day and night temperature, humidity levels, and weather conditions such as snowstorms and thunderstorms. For experiments requiring mountaineering, the skill of the researcher must match the difficulty of the climb. Mountain terrain, particularly the presence of streams in the path of avalanches, can be an environmental factor in an avalanche accident. Topographical features are combined with factors such as avalanche warnings from local weather stations and large amounts of snowfall in the immediate vicinity.

Heat stroke is an environmental factor when experiments are conducted outdoors for prolonged periods in the summer. Particular attention should be paid to environments with high temperatures and humidity and strong sunlight, requiring appropriate breaks and hydration. In some countries, meteorological services have announced the weather conditions that make people more susceptible to heat stroke.

### 5.4.4 Central Liveware/Operator

Central liveware (Lc) is an accident-contributing factor, usually for the operator who triggered the accident or incident. Applying the concepts of the socio-technical systems model developed by Grech et al. (2008) to accidents in universities, etc., the situation of the person at the time of the accident can be divided into six categories: sensory limitations, physiological factors, pathological factors, psychological factors, workload management and knowledge, skills, and education/training. When conducting an accident investigation, starting with the Lc at the scene of the accident will help clarify the causal factors that led to the accident.

#### 5.4.4.1 Sensory Limitations

Sensory limitations result from each process of human information processing. Sensory limitations are one reason why accidents occur when chemicals are mistakenly taken from containers of the same color and shape, which are indistinguishable from each other.

There are various models and theories of the information processing process, but generally, five processes are recognized: (1) sensing, (2) perception, (3) decision, (4) action, and (5) feedback (Hawkins 1987; Endsley 2000; Yokomizo and Komatsubara 2013).

The process of human information processing becomes clearer when Endsley's situational awareness model is used. According to Endsley (2000), situation awareness consists of perception, comprehension, and projection. In the process of perception, the sensory organs receive a variety of information that is relevant to the operator's goals and exceeds thresholds. Comprehension involves the coordination and integration of information from the operator's own situation and the situation around the operator, resulting in an understanding of the situation that is consistent with the operator's goals. Projection involves predicting future events and dynamics from current events and dynamics. Based on the projection, the operator makes a decision. This is followed by the execution of actions and feedback as a series of processes. With situation awareness, if the time between perception and decision is short, the operator may be forced to make poor decisions on the spot, leading to accidents. Situational awareness, decision, and action are influenced by the operator's goals and objectives, memory, experience, training, stress, workload, etc.

Endsley's situational awareness diagram can be illustrated in everyday life, e.g. driving a car, as follows. You are driving your car on a narrow road in a certain town, and you see a bicycle in the distance. There are no other vehicles accompanying or opposing you. This is a perception. As you approach the bicycle, you realize that the man-driven bicycle is moving slowly along the edge of the narrow road in the same direction as your car. This process is called comprehension. You consider that if a car drives very close to the cyclist at the current speed, a high speed, the cyclist may wander off and come into contact with your car, potentially causing a collision. If you drive slowly, you can stop and avoid a collision with the cyclist, even if the cyclist comes toward the center of the road. This process is known as projection. You then decide to drive slower when approaching and overtaking the cyclist. This is the decision process. Based on this decision, you drive slowly, keeping an eye on the cyclist's behavior and keeping as much distance as possible from the cyclist as you approach him. This is the action process. You check in your rearview mirror to see if your actions have allowed you to safely overtake the cyclist. This final process is feedback.

Figure 5.1 shows an accident investigation and analysis of a collision between two cargo vessels at a narrow channel crossing (Japan Transport Safety Board 2014). The captain's situational awareness and the decisions and actions leading up to the collision is presented and visualized using Endsley's situational awareness model. The collision occurred only four minutes after the captain of the cargo vessel Kota Duta (hereafter referred to as "Captain A") sighted another vessel. Captain A of the cargo vessel Kota Duta (hereafter referred to as "vessel A") was aware that another vessel, Tanya Karpinskaya (hereafter referred to as "vessel B"), was approaching vessel A and could collide at a narrow channel crossing. It was clear that Captain A's situation awareness was reasonable and that Captain A did nothing wrong. However, the process of making a decision and taking action as to how vessel A would specifically alter its course to avoid the collision with vessel B took time, and the information received from Captain B on how to maneuver to avoid a collision was inadequate. Captains A and B communicated by Very High Frequency (VHF) radio and agreed that both vessels would turn left to avoid the collision. Captain A assumed that vessel B's collision

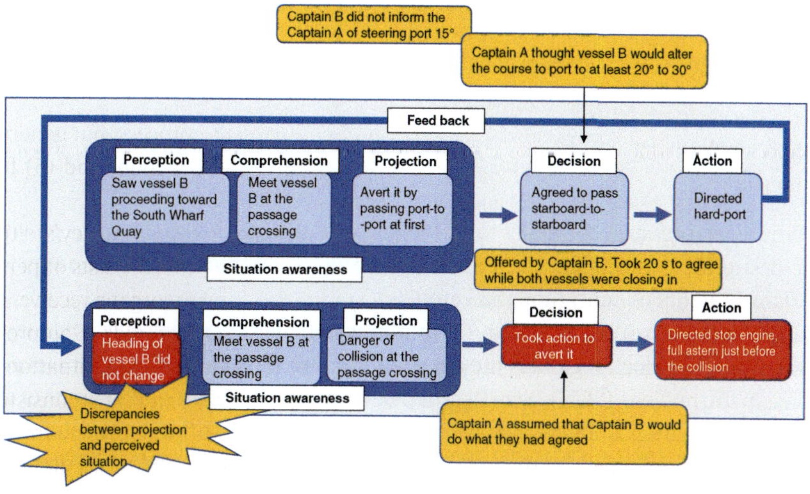


**Figure 5.1** Information process of Captain A in a collision case (Fukuoka 2019). *Note:* This diagram shows the information process of Captain A in relation to the collision accident in the narrow channel using Endsley's situational awareness diagram. The diagram shows the chronological sequence of events that led to the situation not developing as Captain A had originally envisaged.

avoidance maneuver would be the same as that of vessel A, i.e. hard port or 35° to the left, and then initiated a hard port turn to the left. However, during feedback, Captain A becomes aware that vessel B was not turning left and that there was a risk of collision. Until shortly before the collision, Captain A believed that the agreement with Captain B by VHF had been honored and that a collision could be avoided. Therefore, he did not take any action to avoid the collision, such as putting the engine full astern, until just before the collision. The same was true for Captain B, who believed in the agreement and did not take any collision avoidance measures immediately after sighting vessel A. The two vessels collided at the channel crossing and vessel B sank.

#### 5.4.4.2 Physiological Factors

Physiological factors relate to the individual as a complex organism made up of diverse human systems. These include health, alcohol, drugs, nutrition, stress, and fatigue. Fatigue is a major accident-contributing factor in the Lc.

##### *5.4.4.2.1 Accumulation of Fatigue*

Fatigue is often confused with sleepiness or tiredness. While these can be symptoms of fatigue, it is possible to be dangerously fatigued without feeling sleepy or tired. Fatigue is defined as a reduction in physical and/or mental capacity as a result of physical, mental, or emotional exertion that adversely affects almost all physical abilities, including strength, speed, reaction time, coordination, decision-making, and balance (IMO 2001).

In the aviation industry, fatigue contributes to 4–7% of commercial aircraft accidents and incidents and almost 95% of fatigue is attributed to inadequate amounts of sleep (Caldwell and Caldwell 2003).

##### *5.4.4.2.2 Factors of Fatigue*

Fatigue is associated with physiological factors, such as slow brain activity, changes in eye movements, and pupil response. Biological indicators of fatigue have not been identified but have been associated with alertness level (Moore-Ede 1993). Alertness level is measured by electroencephalography, electrooculography, and electromyography. There are three main factors that influence alertness level (Caldwell and Caldwell 2003).

- Homeostatic sleep pressure
- Circadian mechanism
- Sleep inertia

Homeostatic sleep pressure correlates with the amount of time awake since the previous sleep period. The desire for homeostatic sleep after awakening is low but becomes more pronounced after more than 16 hours of continuous wakefulness.

The circadian mechanism is a circadian rhythm in which alertness level acts as a daily cycle of sleep and the biological clock regulates the circadian rhythm. The biological clock is regulated by the suprachiasmatic nucleus (SCN), a small group of neurons near the hypothalamus of the brain. Light and dark information from the eyes is transmitted to the SCN through the subthalamic tract to the hypothalamus, synchronizing our body rhythms to a 24-hour cycle (Moore-Ede 1993). When homeostatic and circadian mechanisms overlap late at night or at dawn, alertness levels drop to dangerous levels.

Sleep inertia is a brief period of drowsiness immediately after awakening.

In addition to these three factors causing fatigue, accumulation of fatigue, poor sleep quality, and working early in the morning are also relevant. Accumulation of fatigue is formed by the sleep debt accumulated from insufficient sleep in the previous 24 hours. There is an existing sleep debt inherited from insufficient sleep the previous day. For most people, an adequate amount of sleep requires seven to eight consecutive hours, although the amount of adequate sleep decreases with age. Studies have shown conflicting results, with recovery from fatigue taking several days to three weeks (Nishino 2017).

Sleep quality is reduced by environmental factors, circadian mechanisms, sleep disturbances, and psychological stress. The timing of sleep also affects its quality. Night and early morning shift workers work with these factors. They force their bodies to wake up during the night shift and then sleep during the day after the shift. According to Caldwell and Caldwell (2003), night shift workers slept less than five hours between 11:00 and 19:00. Waking up after three to four hours of sleep makes it difficult for most night shift workers to sleep during the day. Night shifts can lead to a vicious cycle of accumulated fatigue, and unless something is done about it, there is no time to recover from the accumulation of fatigue.

#### 5.4.4.2.3 *Cases Where Fatigue Was the Cause of the Accident*

According to the National Transportation Safety Board (NTSB), the accident that occurred at 00:09 on 24 March 1989, when the oil tanker *EXXON VALDEZ* ran aground on Bligh Reef in the Gulf of Alaska, spilling a large quantity of crude oil into the surrounding waters, was the first case in which a national accident investigation agency identified fatigue of the third officer in command of the ship at the time of the accident as a probable cause of a marine casualty (NTSB 1990).

#### 5.4.4.2.4 *Impact of Fatigue and Fatigue Mitigation Measures*

The case of the *EXXON VALDEZ* provided some insight into the typical effects of fatigue. The specific effects are listed below:

- Reduced accuracy and timing
- Reduced attention span
- Unconsciously falling asleep
- Unconsciously accepting lower standards of performance
- Loss of ability to integrate information
- Reduced ability to grasp the big picture
- Reduced social interaction

The reduction in alertness level because of fatigue is due to physiological limitations that cannot be overcome by training or motivation. Among the issues discussed by Caldwell and Caldwell (2003) in relation to countering fatigue, some effective measures to manage alertness and fatigue, which are applicable to researchers and students on night shifts or in sleep-restricted conditions include the following:

- Reduce the mental demands of the night shift as alertness drops between 01:00 and 06:00.
- Encourage night shifts to double-check everything.
- Use the biological clock drop point when taking breaks. Choose a break or nap time that coincides with a drop in alertness level, such as after lunch, before dawn, or late at night.
- Take strategic naps to combat sleep deprivation. Even short naps of 20–30 minutes can improve productivity in sleep-deprived people when extended periods of uninterrupted sleep, whether during the day or night, are not possible. Effective naps should be taken immediately after the start of a long working day or at appropriate times to match the internal biological clock.

- Try to get a comfortable night's or day's sleep. Often a change of environment does not allow the same quality of sleep as at home. Try to make the away environment as homely as possible, for example, by bringing mementos from home, such as your own pillow or family photos. Using a sleep mask can help block out light. Foam earplugs can be used to minimize noise and disturbance.
- Consider circadian adaptation. If the shift is longer than three days, the shift should be forward or clockwise. This schedule is consistent with the natural ability of humans to easily extend their internal day. If the shift is only one to three days, maintain your daily eating and sleeping habits at home. As it is difficult for the biological clock to adapt to the environmental clock in the short term.

#### 5.4.4.3 Pathological Factors

The presence of a preexisting illness in the experimenter may have a significant impact on the safety of the experiment. In particular, individuals with cardiac diseases, angina pectoris, diabetes, and respiratory diseases, such as asthma, are considered contraindications to scuba diving (Lippmann et al. 2016). Researchers conducting fieldwork activities, not just diving, should check for relevant diseases in advance and consult with an occupational physician, family doctor, or a doctor familiar with diving about whether or not fieldwork or diving activities are permitted.

It is recommended that individuals engage in experiments only when they are in good health. In the event that an individual becomes unwell, it is advised that they refrain from participating in experiments. It is also advised that researchers who are allergic to specific substances investigate in advance whether the materials to be handled in the experiment, the material of the PPE, etc. are allergenic. They should also discuss the experiment with their industrial physician or family physician and take measures to prevent allergic reactions.

#### 5.4.4.4 Psychological Factor (Psychological Pressure)

A characteristic psychological factor for operators in accidents that have occurred at universities, etc. is the psychological pressure caused by the deadline for submission of papers and the approaching end of the experimental period during fieldwork. There have been cases where accidents have occurred as a result of the psychological pressure caused by experimental plans that are difficult to change and allow the researchers not to take more time to safety issues such as risk assessments that should be carried out over time.

#### 5.4.4.5 Psychological Factor (Familiarity, Complacency, Assumptions)

The CSB (2010), which published the accident investigation report on the explosion that occurred during a chemical experiment at a university, stated the following:

*In general, researchers become less correctly aware of hazards as Complacency becomes predominant, as they become more familiar with the task and as their performance becomes more skill-based.*

It is complacency, not familiarity or habituation, that the CSB is concerned with. Although the title of this subject uses the easily understood generic term "familiarity," the technical term in the safety field is complacency, so complacency will be discussed.

Accident investigation reports from other industries often cite complacency as the main factor that led the operator to take the action that caused the accident. The mechanism of complacency is as follows.

When the operator encountered a hazard, such as an explosion hazard, in an experiment before the accident occurred, he omitted some of the risk assessment steps and procedures but was able to achieve his objectives safely without an accident. This event is recorded in the operator's memory as a success story and as long-term memory, which lasts for a long time, rather than short-term memory, which disappears after a short time. The operator then retrieves the memory of past experience when faced with a similar hazard and short circuits it, bypassing the formal risk assessment and procedures as before.

When the author interviewed operators during the accident investigation process, a common statement heard was "I knew the procedures" and "I knew the rules." Even though they had been trained in the rules and procedures, the rules and procedures were overridden by complacency in the field.

In the marine sector, there are studies on the impact of complacency (Fukuoka 2015). An analysis of 28 collisions involving 56 vessels in Japanese waters between 2008 and 2014 found that 52 vessels were operating in improper maneuvers. Several contributing factors were found to be simultaneously present in the occurrence of accidents, of which psychological factors accounted for 89%, mostly due to complacency. The study shows that operator's complacency tends to interfere with the process of collision avoidance maneuvering, resulting in deviations from the COLREGs. In this way, complacency can have a significant impact on the decision-making of operators, who are influenced by their past and recent experiences, which is often associated with accidents.

##### 5.4.4.5.1 Measures of Dealing with Complacency

One method of preventing accidents through compliance is to appoint a person to monitor and review the safety of experiments and operations. In aviation, the majority of accidents occur in the three minutes after takeoff and occur in the eight minutes before landing, for a total of 11 minutes. These 11 minutes are known as the "critical 11 minutes" because of the high probability of accidents occurring. This period is due to the change from automatic to manual operation compared to flying at higher altitudes and the aircraft being more vulnerable to sudden external factors such as strong winds and other weather conditions, bird strikes, and other unexpected events.

The aircraft is operated by a qualified crew of two or three people, usually consisting of a pilot and a copilot. When a pilot is flying an aircraft as pilot flying (PF), a copilot sitting next to him as pilot monitoring monitors the pilot's flight situation using instruments, visual checks, etc. to double check for errors or mistakes. If an error or mistake is found, it is corrected. The reason why the copilot can perform the role of monitoring and checking the pilot is that the copilot has completed the necessary training as an aircraft pilot and has mastered the skills of aircraft operation just like the pilot. The system is designed to prevent accidents before they occur, thus preventing the pilot from making mistakes or errors due to complacency.

A similar approach is being implemented in the marine sector as part of bridge resource management (BRM). BRM differs from standard operating procedures in that it is a method implemented on the bridge or control room by a team of crew members to ensure the safe navigation of the vessel. The concept and understanding behind the method are that the safety of the vessel is not dependent on the decisions of a single person in command, such as the master, but that all decisions and the movements of the vessel as a result of those decisions must be monitored by the crew on the bridge. If the crew believes that a decision taken by the master or the navigation officer in charge of the ship affects the safety of the ship, the crew can challenge the decision. In other words, it is a management technique that ensures the safety of ship operations with a bridge team consisting of the master, navigator, helmsman, and lookout (Swift 2000). The theory behind this, which is also related to the Swiss cheese model, is that BRM creates layers of defense so that errors by the master do not directly lead to accidents. BRM evolved from cockpit resource management in aviation, which maximizes the use of resources in the local workplace.

In universities, etc., laboratory safety is trained to ensure safety through risk assessment, which consists of three processes: hazard identification, risk estimation, and risk evaluation; the monitoring and review process is not required. An analysis of accidents at universities shows that many occur when experiments are carried out alone. As discussed in Chapter 4, OSHA does not recommend that experiments be conducted alone. In light of this, it would be feasible for universities, etc. to conduct experiments with more than one person and have another researcher present who could take on the monitoring and reviewing role, such as a trained researcher, supervisor, etc. In marine experiments, when scuba diving, it would be possible for a diving safety officer or a qualified person who has completed training in scientific diving to perform the role of monitoring and review.

#### 5.4.4.6 Workload Management

Research is often carried out in teams, particularly in the case of experiments involving hazardous chemicals and substances or fieldwork involving scuba diving, where the workload is concentrated on a particular researcher, leading to overload and accidents due to unintentional omission of parts of the work procedures. The team leader or supervisor needs to manage the workload of the team members.

#### 5.4.4.7 Knowledge, Skills, Education, and Training (Conducting Risk Assessment, Including Hazard Identification)

Risk assessment starts with hazard identification, i.e. identifying what constitutes a hazard. Therefore, in the case of chemical experiments, reference materials such as safety data sheets (SDS) and laboratory chemical safety summaries (LCSS) can be used to characterize the chemical of interest. If a procedure for experimenting with a known chemical has been established, the experiment will be carried out according to that procedure. In real laboratories, experiments are often conducted in an environment that differs from the written procedures for a particular experiment, for example, where several experiments are running simultaneously. In such situations, a risk assessment must also be carried out for the environmental conditions.

It was mentioned earlier that risk estimation consists of a severity scale and a frequency scale. If the chemical used in the experiment is a known substance and the human casualties in the event of an accident are low, the severity scale is likely to be low. However, if the substance is unknown and there is a physical hazard, such as the risk of explosion, the severity scale is likely to be high. For more information on the frequency scale, see Section 5.4.1 in this chapter. Following this process, a risk evaluation is carried out to determine whether the risk is acceptable or tolerable. If the risk level is judged unacceptable, the hazard identification, risk estimation, risk assessment, and risk reduction processes are repeated. The risk level should be reduced until it reaches an acceptable or tolerable risk (ISO/IEC 1999). The decision should not be made alone but in consultation with the faculty, PI, supervisor, etc.

Risk reduction measures may be listed in the procedure manual, but risk reduction must be carried out in accordance with the priorities of the hierarchy of control (see Chapter 4). When handling hazardous chemicals, such as explosives, blast shields, explosion-proof film, etc., should be provided and measures taken in advance to mitigate damage in the event of an explosion. If there is a report on the investigation and analysis of accidents that have occurred in the same experiment in the past, it will be possible to know at what process of the experiment the accident occurred and what the accident-contributing factors were. In addition, if specific accident prevention methods are listed, it will be possible to refer to these accident prevention measures. Before carrying out a risk assessment, it is advisable to discuss the method of execution with the supervisor, PI, and confirm the measures that will ensure safety. As part of the risk assessment process, the experiment should prepare a contingency plan in the event of an accident.

In fieldwork, when carrying out the hazard identification process, the surrounding conditions, especially the elements related to the environment of the SHEL model, are very different from the normally stable laboratory environment. The environment that may be encountered during the duration of the fieldwork experiment, the presence of animals and marine organisms that may cause harm to humans, the topography of the seabed and the mountainous terrain, weather, and oceanic conditions must be taken into account. Factual information on previous accidents and their contributing factors in the experimental area of the fieldwork activity under consideration provides an important opportunity to gain prior knowledge of the hazards and potential accident-contributing factors, as well as providing reference material for risk estimation and risk evaluation.

When carrying out fieldwork, it is important to be aware that environmental factors can have a significant impact on safety and to establish criteria for stopping experiments when weather and sea conditions reach a certain level. In contrast to university campus experiments, it is not possible to assume that rescue or medical organizations will be able to reach the scene of an accident within a short time in fieldwork, even if they are called by telephone. Therefore, the conditions for starting an experiment are that the emergency response plan and training have already been completed, including the preparation of automated external defibrillators, cardiopulmonary resuscitation, and first aid kits, e.g. atmospheric oxygen inhalation for minor cases of decompression sickness, and the mastery of first aid measures.

#### 5.4.4.8 Knowledge, Skills, Education, and Training (Education and Training)

Regardless of the field, education and training are frequently pointed out as accident-contributing factors as a result of accident investigations. The substance of the point raised is that there are two types of SOPs: general SOPs and specific SOPs, and according to accident investigations at universities, etc., the chemical hygiene plan (CHP), which is a general SOP, exists. However, there are no specific SOPs, which are tailored to the laboratory and based on risk assessments of specific chemicals (CSB 2010). OSHA has responded to this challenge identified by the CSB by recommending that university chemical laboratories conduct risk assessments for laboratory-specific hazards and provide education and training (OSHA 2013).

There have been cases where accidents have occurred because the researcher had knowledge of the explosion hazards of certain chemicals but had never received practical training on how to reduce the risk in specific ways. It should be noted that having knowledge and retaining skills as accident-contributing factors are very different in terms of accident prevention.

Accidents at sea and in the mountains have been observed where researchers and students have had accidents because they were not adapted to the level of difficulty of diving and climbing. A series of educational and training programs are required, especially for fieldwork in the sea or mountains, to increase the level of training according to the difficulty of the skills required, such as the depth of the sea area where experiments are to be conducted, the topography of the mountains, and the equipment to be used.

### 5.4.5 Peripheral Liveware

Lp mainly refers to the relationship between individuals and the people in the organization to which they belong and is broadly divided into laboratory safety management, university safety management, and relevant organizations that are involved in accidents.

Laboratory safety management is subdivided into education and training, communication, teamwork, experiment-specific risk management, and accident/near miss handling, which are part of Lc.

Safety management at universities is divided into the following categories: setting standards for suspension of experiments and work, risk reduction measures based on risk assessment, approval of changes to experimental plans, maintenance and inspection systems for equipment, systematic education and training programs, workplace inspections, audits and corrective actions, systems for conducting investigations in the event of accidents and near misses, operation of an organizational emergency response plan, EHS and laboratory coordination, commitment of university top management, and an organizational structure that ensures safety.

Accident-contributing factors present in external organizations are included in the Lp because accidents involve other nonuniversity organizations related to the experiment.

The CSB's accident investigation analysis of explosion accidents at chemical facilities and universities has identified a number of accidents contributing factors in the Lp. As explained in accident models in Chapter 3, safety recommendations are not directed at individuals, but at the organization that has fostered the environment in which individuals make mistakes. It is therefore important to understand and elucidate the organizational factors during the accident investigation and analysis process.

#### 5.4.5.1 Communication

Communication is part of laboratory safety management and refers to the exchange of safety-related information among researchers, between researchers and faculty, and among technical staff. Poor communication is often a contributing factor to accidents.

Poor communication includes cases where there is no opportunity to share safety information, such as procedural problems or near misses that occur during experiments, and where safety information is not discussed and communicated with the faculty or principal investigator (PI) in the laboratory. In addition, experiment notes and group meetings are devoted solely to matters relating to the planning and progress of experiments, and there is no sharing of safety information, for example.

Technical staff responsible for the care of marine organisms and animals in the laboratory may encounter near misses during daily feeding operations, but in some cases safety information from technical staff is not communicated to researchers and faculty, and a system for sharing information across the laboratory needs to be developed.

Communication has four elements: sender, receiver, message, and medium (Parrott 2011). Face-to-face communication is most effective because the sender and receiver can use tone, intonation, and body language to emphasize the intended meaning. Modes of communication using devices, such as radio, mobile phones, and intercoms, require a great deal of attention to be accurate to the other party. It is necessary to speak accurately, use standard language, express clearly, listen carefully, check the message, and ensure that the words of both parties do not overlap.

These modes cannot be applied to underwater communication in fieldwork. In experiments involving scuba diving, when an unusual situation occurs underwater, simple conversation is carried out using hand signs and other means, as the tools for communicating with a buddy have not yet been developed. In some cases, fieldwork involving scuba diving could have prevented the worst if sufficient underwater communication had been available.

In other industries, closed communication loops are used to prevent accidents caused by a lack of communication. This is a method of confirming that the message sent was clearly understood by the recipient and that the action was achieved as intended by the sender. For example, when the master gives a steering instruction to the helmsman, the helmsman recites the instruction and sets the rudder angle as indicated by the steering gear. When the helmsman acknowledges the indicated rudder angle, he shall report to the master that the indicated rudder angle has been reached. The master shall confirm to the helmsman that the verbal instruction has been achieved as an action and issue a message.

#### 5.4.5.2 Teamwork

Research is usually carried out by a team of researchers, students and others. Teamwork is particularly important in fieldwork to ensure safety. Teamwork means that responsibility and authority are clear and that everyone fulfills his or her role. A team is a group of people working together toward a common goal. Teams have a leader but are most effective when they fulfill the role of both nominated and de facto leader. Each member of the team then needs to know his or her role and the roles of the other members. As a result, a common mental model is formed within the team, and team members come to understand that different tasks are a means to achieve their goals. Team members have a responsibility to tell the leader if they think the leader is making a mistake (Adams 2006; Swift 2000).

#### 5.4.5.3 Experiment-specific Risk Control

In each laboratory, a wide variety of experiments is carried out simultaneously by faculty, staff, and students. General guidelines or general SOPs, including the Chemical Hygiene Plan developed and maintained by the administrative department, are not sufficient to ensure the safety of a wide variety of experiments. For this reason, each laboratory must formulate specific SOPs that address the hazards specific to the experiment and cover areas not listed in the general SOPs.

Specific SOPs are outlined in Section 5.4.4.7 and are necessary when dealing with new materials, unknown chemicals, new processes, etc. There are no precedents, no previous studies, and no proven procedures or protocols. In such cases, as recommended by the National Academy of Sciences (2011), the experience of the laboratory personnel and the level of potential hazards associated with the experiment should be considered, and the supervisor and safety professionals should be consulted before conducting a risk assessment. In addition, according to OSHA (2013), the risk assessment should be reviewed with a chemist who has experience with similar materials and protocols.

Risk reduction based on the hierarchy of control does not mean that the residual risk is zero. Particularly in experiments where there is a possibility of physical hazards such as fire or explosion, it is necessary to mitigate the damage by preparing an emergency response plan.

#### 5.4.5.4 Accidents and Near-miss Response

If an accident or near miss occurs, the laboratory must communicate the facts and factors contributing to the accident or near miss within the laboratory and work to prevent a recurrence. As is clear from accident investigations and analyses in other industries, if a near miss occurs during an experiment but the contributing factors that led to the near miss are not investigated and analyzed, and the experiment is continued in the same way, there is a high probability that an accident will occur. For a discussion of the theory of accidents and near misses, see Section 2.5 in Chapter 2.

#### 5.4.5.5 Establishment of Criteria for Suspension of Experiments and Work

In the safety risk management and safety assurance processes that make up the SMS, risk assessment is carried out prior to the start of an experiment, and the experiment or work is not started until sufficient risk reduction has been achieved through engineering and administrative controls and the risk is deemed acceptable. However, if during the experiment the risk controls that were the basis for determining the risk to be acceptable are found to be ineffective, or if new risks arise, the experiment must be stopped and the risk assessment carried out again. The department responsible for EHS in universities, etc. must consult with the faculty in advance to establish criteria for stopping an experiment if an accident or serious near miss occurs in an unsafe situation. For more information, see Section 2.6.4 in Chapter 2.

#### 5.4.5.6 Development and Operation of General SOPs and Experiment-specific SOPs

The S in SOP stands for standard and refers to the arrangements for the content and working methods of a product or process. Standardization is the process of standardizing processes in other industries, and the reason for standardizing processes is to minimize process variability and stabilize the quality of products and services (Yamada 2006).

Standardizing the way experiments are conducted can improve safety. That is, standardization ensures that the process and results of the intended experiments are always the same, even if the researchers conducting the experiments are different. As part of administrative control, it was mentioned earlier that SOPs need to include general and specific SOPs. The EHS needs to develop a general SOP that applies to the whole university and department and keep it up to date by monitoring and reviewing its implementation. The general SOP is a common procedure document for laboratories throughout the university, etc. As there is a wide variety of research topics, there will be ones that are not covered by the general SOPs. Specific SOPs are developed by individual laboratories for research topics that cannot be covered by the general SOPs. For example, a specific SOP is a procedure that applies to a particular substance when handling a chemical in an experiment. Therefore, a specific SOP is a procedure or protocol that applies to a specific experiment and should be developed in collaboration with other researchers, including those outside the university who are conducting research on the same topic, and experts in the field, especially if new and unknown materials are the subject of the experiment.

#### 5.4.5.7 Approval of Changes to Experimental Plans

During the course of an experiment, whether it is a chemical experiment or fieldwork involving scuba diving, circumstances may arise that require changes to an already approved experimental plan. In such cases, the official procedure for changing the original experimental plan will be followed. The changes to the experimental plan must be assessed for their impact on the approved risk controls, and a new risk assessment has to be carried out for the substances, equipment, and procedures to be changed. The changes must be approved by the body that approved the original experimental plan, such as the Health and Safety Committee. In universities, etc., where a SMS is planned to be established, a documented procedure or approval process for changes to the plan needs to be specified.

#### 5.4.5.8 System for Maintenance and Inspection of Equipment, Etc.

The administrative offices, in particular the EHS and the facilities department, may carry out maintenance and inspection of research facilities, equipment, fire protection, and extinguishing systems, etc. as required by regulation for the university as a whole, or laboratories may carry out maintenance and inspection of equipment and instruments on their own initiative. Maintenance and inspections include checking laboratory equipment such as flasks, beakers, and reaction vessels for cracks; bonding and earthing equipment required when handling flammable substances, gas detectors, and fire alarms. These statutory and voluntary maintenance checks must be communicated from the administrative office to the laboratories concerned and carried out as part of the SMS.

#### 5.4.5.9 Systematic Education and Training Programs

An issue often highlighted in detailed accident investigation reports is the lack of a systematic education and training program. In particular, accidents have occurred in experiments with hazardous chemicals in laboratories and in fieldwork involving scuba diving because there is a large gap between the skills required to carry out the experiment and the skills of the researchers to carry out risk assessments and scuba diving. To improve the skills of researchers, universities as an organization need to provide systematic, long-term education and training programs. This program requires a system that allows researchers to acquire the skills required for their field of research by the time they start their experiments and allows the head of the laboratory and the EHS to check that the researchers have mastered the skills.

#### 5.4.5.10 Workplace Inspections, Audits, and Corrective Actions

The importance of workplace inspections and audits is that the work practices and procedures to be introduced into the organization are developed in the Plan process, trained in the Do process and then implemented in the local workplace in accordance with the training. The Check process is to check that the SMS is being operated correctly and that any issues identified as requiring improvement are being addressed and safety is being further improved. If any safety-related problems are found as a result of workplace inspections or audits, the laboratory will be instructed to take measures to rectify the problems, and the inspector or EHS will follow up to ensure that the issues raised are addressed on-site. For details, see the core of the SMS inspection or verification (check) in Chapter 4.

#### 5.4.5.11 Implementation Structure for Conducting Accident and Near-miss Investigations

The structure for conducting accident and near-miss investigations is as follows: when a near miss that could lead to an accident or serious accident occurs during an experiment, a department that systematically investigates accidents and near misses, or a third-party organization contracted by universities, etc. will conduct a comprehensive accident and near-miss investigation based on the evidence. By conducting a detailed accident and near-miss investigation, they will clarify the contributing factors that led to the accident or near miss and will make safety recommendations to the laboratory that caused the accident, identifying safety-related issues that need to be improved to prevent a recurrence.

It also refers to a system in which the EHS can contribute to the prevention of accidents by providing safety recommendations to all laboratories involved in the area where the accident occurred as a lesson learned.

The response to accidents and near misses in universities has been largely unprepared due to the fact that universities, etc. have a different organizational structure from industry and have not adopted safety management regulations as university rules, which has delayed the introduction of SMS and the recognition of its effectiveness. In many universities, when accidents occur during experiments, accident investigations are carried out by the faculty and staff of the laboratory concerned using their own investigation and analysis methods, resulting in a lack of uniformity and reliability in accident investigation and analysis.

#### 5.4.5.12 Implementation Structure for Emergency Response Plans

An emergency response plan is a mitigation measure in the event of an accident or disaster. According to the National Academy of Sciences (2011), emergency response plans for chemical handling are described as something that should be developed as part of the risk assessment process. Laboratories and local workplaces must have emergency response plans for dealing with fires, explosions, chemical spills, injuries, etc., and researchers and students must be trained in advance.

#### 5.4.5.13 Coordination Between EHS and the Laboratory

The lack of cooperation between the safety inspector and the PI or faculty during workplace inspections, and cases where workplace inspections are conducted in the absence of faculty, have been noted in accident investigation reports (CSB 2010; UC Center for Laboratory Safety 2016). As a result of this lack of cooperation, the safety inspectors are unable to obtain information such as the chemicals used in the experiments conducted in the laboratory, the procedures, and the measures to be taken in the event of an emergency, while the faculty is unable to obtain accident data and safety-related information held by the safety inspectors. The safety inspectors will inspect safety in a situation awareness, where they have no knowledge of the situation in the laboratory. Therefore, they cannot inspect the handling of risks in accordance with the actual experiment. Cooperation between the two sides is necessary from the point of view of ensuring laboratory safety.

#### 5.4.5.14 Commitment of University Executives

Decision-making power in the organization rests with top management, and the policy and commitment of top management to EHS are important. In universities, etc., raising awareness of the importance of the environmental, safety, and health fields through a top-down approach provides an opportunity for faculty, staff, and students in laboratories to recognize that it is an important issue for the whole university and can influence behavior in laboratories. It would also address the issues of staffing, funding, equipment, and training to implement the policy. The powers and responsibilities of individuals or departments involved in EHS, such as the President, Provost, Vice President, Heads of Department, and the EHS, should be clearly defined, and each should be aware of their responsibilities and authority.

#### 5.4.5.15 Organizational Structures to Ensure Safety

A comprehensive accident investigation will uncover a number of accident-contributing factors. If these accident-contributing factors are not corrected as soon as possible, a similar accident may recur in the same organization. If this situation occurs, measures should be taken to stop experiments and field research activities until the accident-contributing factors are rectified in order to prevent similar accidents from occurring.

The CSB (2010), in its accident investigation report into the explosion at University A, noted that the EHS conducting workplace inspections should be able to report directly to individuals or departments with organizational authority to implement the corrective actions identified. Specifically, it was the Dean, and the Provost or the Vice President for Research who could directly influence faculty and PIs regarding research safety at University A. The EHS belonged to the facilities office and was under the direction of the vice president for administration and finance, which meant that it could not directly influence faculty or PIs. Therefore, it was recommended that the structure be changed to allow the EHS to report directly to those individuals or departments.

If the department in charge of accident investigations on campus is a research oversight unit or a unit reporting directly to the president and has the authority to suspend experiments based on the criteria for suspending experiments, it can implement the suspension of experiments.

Universities have different organizational structures, but from an accident prevention perspective, the organizational structure needs to be changed so that in the event of a near miss resulting in a serious accident, a specific department has the authority to carry out a thorough accident investigation without hesitation and to suspend experiments similar to the accident until a situation is reached where safety can be ensured.

#### 5.4.5.16 Involvement and Improvement of Other Institutions

Analyses of accident investigations in other industries have shown that the operations, laws, regulations, and policies of organizations, including international organizations and government agencies, outside the framework of the organization in which the accident occurred are involved as factors contributing to the accident (Fukuoka 2019). In the case of accidents in universities, etc., the safety requirements of research funding bodies, national safety laws and regulations, the design and the hierarchy of controls by manufacturers of facilities and research equipment, and the licensing requirements of private certification bodies are all relevant.

## 5.5 Safety Culture

Safety culture as a contributing factor to accidents is not discussed in this chapter. According to Reason (1997), safety culture consists of four subcomponents: reporting culture, just culture, flexible culture, and learning culture. These four subcomponents work together to create an informed culture, which is equivalent to the safety culture. Reporting culture is the willingness of people to report their own errors, near misses, and hazardous situations to the appropriate department. A just culture is one in which the provision of safety-related information found in each department is encouraged, while acceptable and unacceptable behavior is defined. A flexible culture is one in which the organizational structure shifts from a hierarchical mode to a flatter, more specialized structure in the event of an emergency, with local workplace experts taking the lead. A learning culture means that the organization draws on lessons learned and implements reforms where necessary.

In other words, a safety culture can be achieved by universities, laboratories, and individuals recognizing each of the SHEL elements listed in Sections 5.4.1 to 5.4.5 of this chapter and acting to ensure that safety is maintained in their day-to-day teaching and research activities. To put it another way, an organization can be said to have a safety culture when all the contributing factors in Table 5.1 are shared by individuals and the organization, and the SMS is in place to improve problem areas.

It should be noted that IMO (2003) defines safety culture as "a culture in which significant efforts are made to acquire the knowledge to reduce substantially the risks to individuals, ships and the maritime environment."

**Table 5.1** *(Continued)*

| SHEL element | Broad category | Subdivision | Accident-contributing factors | Relevant higher level accident-contributing factors (subject to safety recommendations) |
|---|---|---|---|---|
| | | Physiological factor (circadian rhythms and alertness) | The experiment was conducted during the late night and early dawn, a period of reduced attention due to circadian rhythms | Review of experimental plan |
| | | Pathological factors | The researchers had preexisting medical conditions that affected their ability to carry out the experiments (including diving and other fieldwork activities) | Regular health check-ups (including statutory) → development and operation of general SOPs and experiment-specific SOPs |
| | | Psychological factor (psychological pressure) | Researchers were under psychological pressure, such as not being able to meet the deadline for submitting their dissertation | Review of experimental plan → approval of changes to experimental plan |
| | | Psychological factor (familiarity, complacency, assumptions) | When dealing with potentially hazardous substances/processes, assumptions, distractions, and complacency acted to reduce the level of awareness that they were dangerous | • Avoidance of independent experiments and implementation of monitoring & review<br>• Development and operation of general SOPs and experiment-specific SOPs |
| | | Workload management | Individual workload management was not in place (workloads that cannot be handled by one person) | Review of the team's role assignment. → Approval of changes to the experimental plan |
| | | Knowledge, skills, education, and training (conducting risk assessment, including hazard identification) | (Conducting risk assessment, including hazard awareness)<br>• They were not aware of the hazards of hazardous materials and potentially hazardous chemicals, including substances used in experiments and operations, explosive mixtures<br>• They did not carry out a risk assessment of the location where the fieldwork was to be conducted | Formulation and operation of systematic education and training programs |
| | | Knowledge, skills, education and training (education and training) | • General safety training for experiments and hazardous substance-specific training was not provided<br>• Training specific to hazardous chemicals or experiments was not provided | Formulation and operation of systematic education and training programs |

| | | | | |
|---|---|---|---|---|
| Peripheral liveware | Laboratory safety management | Communication | • Safety information about the experiment was not shared among researchers<br>• The PI/faculty was not consulted when uncertainties arose<br>• Group meetings were not held as a forum for discussion and exchange of information on safety-related issues | Development and operation of experiment-specific SOPs |
| | | Teamwork | • Roles and responsibilities of those involved, including supervisors and accountable persons, were not clearly defined<br>• Documented roles were not implemented | Development and operation of experiment-specific SOPs |
| | | Experiment-specific risk control | • Experiment specific procedures and protocols have not been developed or trained<br>• Risk assessments have been carried out for the facilities and equipment used and for unknown chemicals/processes, but no risk mitigation measures, such as inherently safe design or engineering controls, have been implemented | Development and operation of experiment-specific SOPs |
| | | Accidents and near misses response | • Accidents or near misses occurred in the laboratory or local workplace in the past<br>• These events were not investigated, analyzed and shared with at least relevant laboratory personnel | Implementation structure for conducting accident and near-miss investigations |
| | University safety management | Establishment of criteria for suspension of experiments and work | There were no criteria for stopping experiments/operations when new risks are identified during experiments/operations | |

(Continued)

Table 5.1 *(Continued)*

| SHEL element | Broad category | Subdivision | Accident-contributing factors | Relevant higher level accident-contributing factors (subject to safety recommendations) |
|---|---|---|---|---|
| | | Development and operation of general SOPs and experiment-specific SOPs | General guidelines or general SOPs and experiment-specific SOPs/specific guidelines ware not formulated nor constantly updated through monitoring and review of their implementation | |
| | | Approval of changes to experimental plan | There was not a documented procedure or approval process for changes to the work/experimental plan or process | |
| | | System for maintenance and inspection of equipment, etc. | Legally required maintenance and inspections of research facilities, equipment, fire protection and extinguishing systems, as well as maintenance and inspections of equipment and instruments voluntarily performed by the laboratory, were not carried out | |
| | | Systematic education and training programs | • There was no assessment of researchers' competences, no systematic education and training program<br>• There was no structure to ensure that all new comers received systematic education and training programs | |
| | | Workplace inspections, audits, and corrective actions | • Safety inspections, workplace inspections and audits of laboratories and local workplaces were not carried out on a regular basis<br>• EHS did not follow up and take appropriate action to correct any unsafe conditions found | Organizational structures that ensure safety |
| | | Implementation structure for conducting accident and near-miss investigations | • There was no system for investigating and analyzing near misses and accidents that occurred in the laboratory and field, and for sharing lessons learned across the university<br>• Systems had been put in place but were not working/effective | |

| | | | |
|---|---|---|---|
| | | Implementation structure for emergency response plans | • No emergency response plan (e.g. fire, explosion, chemical spill, death, or injury) had been formulated in the laboratory or local workplace<br>• Plans had been formulated but were not effective as no drills had been conducted |
| | | Coordination between EHS and the laboratory | • There was no coordination between PIs and EHS, and no faculty or PI is present during safety and workplace inspections<br>• It was not made clear in writing that experiments and safety are inseparable |
| | | Commitment of university executives | • The President's commitment to environmental, health and safety (EHS) was not clearly stated in writing<br>• Powers and responsibilities of the President, Provost, Department Heads, and EHS were not clear |
| | | Organizational structure to ensure safety | • If EHS discovers unsafe conditions that were not controlled, etc. during workplace inspections, the authority to stop the experiment was not delegated to a higher authority in EHS<br>• Workplace inspections were carried out by an external contractor and there was no mechanism for reporting unsafe conditions discovered during inspections to the Dean of Research, Provost, or President who has influence over the faculty |
| | External organizations relevant to accident-contributing factors | Involvement and improvement of other institutions | There were problems with the legislation and operational status of the organizations involved in the accident (e.g. local authorities, industry, government agencies, international organizations, etc.) |

*Note:* This table shows the elements of the software, hardware, environment, liveware (SHEL) model with modifications to adapt them to accidents occurring in universities, etc. In the right-hand column of the table, actions to correct accident-contributing factors identified as a result of accident investigations are listed. Although there are many items in the organizational factors that are not listed in the safety recommendations, they are directly subject to the safety recommendations and specific measures to prevent accident-contributing factors will be issued as safety recommendations according to the actual accident situation.

## 5.6 Conclusion

This chapter provides a new perspective of insight into the safe conduct of laboratory experiments and field research activities in marine, mountain, and wilderness environments, as the concept of human factors can be used for accident prevention for faculty, researchers, and students in laboratories working on accident prevention in universities, etc. It also provides specific topics for in-depth accident investigations and analysis for EHS and professional accident investigation teams investigating accidents in universities, etc. It is recommended to be read in conjunction with Chapter 7, "Contributing Factors Found in Accidents."

It was in the second half of the twentieth century that safety researchers and practitioners around the world began to recognize the importance of human factors in accident prevention. During this period, socio-technical systems were evolving, and industrial structures were becoming more complex, as were the factors contributing to accidents in these industries. Until this time, accident investigation and analysis had focused exclusively on human error, but this approach made it impossible to accurately identify the overall picture of accident occurrence and to elucidate the faulty elements. In this environment, there was a need for a concept and accident model that could include the social components. The concept of human factors was developed and modeled in the SHEL model, particularly in the field of aircraft accidents, where accidents occurred frequently.

The objectives of human factors research and analysis with the ultimate aim of improving the safety of aircraft operations are as follows: (1) to clarify the relationship between human behavior and its development into accidents; (2) to identify the safety implications of the limitations of human capabilities and behavior using the sciences such as physiology, psychology, sociology, and ergonomics; and (3) to issue safety recommendations to prevent recurrence and mitigate harm for accidents that occur as a result of incorrect actions or decisions made by individuals or groups.

It was in 1984 that Hawkins published a modified version of the SHEL model developed by Edwards in 1972. At that time, the concept of SMS was underdeveloped, and it was not fully recognized that the SMS could contribute to accident prevention. Therefore, this chapter supplements the Lp of the model by adding the content of SMS and modifies the prototype of the SHEL model by changing the software to reflect the procedures, practices, and emergency responses that were actually in use at the time of the accident in the local workplace. This will clarify how the SMS established by an organization was applied at the local workplace and how the gap between what the SMS required and the actions taken by operators and the research team at the local workplace contributed to the occurrence of the accident. Furthermore, a detailed study and analysis of the interrelationship situation between the human element and the surrounding people and things, including software, hardware, environment, associated organizations, relevant administrative and private institutions, etc., is carried out to enable the identification of problems in the interrelationships.

Software includes regulations, protocols, standard operating procedures, manuals, checklists, charts, maps and publications, essential qualifications and certifications required for the task, guidelines issued by industry bodies, etc. in use at the local workplace at the time of the accident. Hardware includes the human–machine interface, the condition of equipment and instruments related to the accident, and the installation of gas and oxygen concentration detection equipment in the laboratory. Environment includes not only the laboratory but also the outdoor environment such as rivers, seas, and mountains. It includes the relationship between operators and the natural environment, such as environmental factors in the laboratory that can cause fires and explosions, weather and sea conditions, seabed topography, and mountain topography. Central liveware presents the accident-contributing factors of the operator and is divided into six categories: sensory limitations; physiological factors; pathological factors; psychological factors; workload management; and knowledge, skills, and education and training at the time of the accident. Lp refers primarily to the relationship between the operators and the people in the organization to which they belong and is broadly classified into laboratory safety management, university safety management, and involvement in accidents in external organizations.

Laboratory safety management is divided into education and training, communication, teamwork, experiment-specific risk management, and accident/near-miss handling. University safety management is divided into the following categories: establishment of criteria for suspension of experiments and work, development and operation of general SOPs and experiment-specific SOPs, approval of changes to experimental plan, system for maintenance and inspection of equipment, etc., systematic education and training programs, workplace inspections, audits and corrective actions, implementation structure for conducting accident and near-miss investigations, coordination between EHS and the laboratory, commitment of university executives, and organizational structure to ensure safety. As the accident involved organizations other than the university where the experiment was carried out, the factors contributing to the accident and the corrective actions taken by these organizations have been included in the Lp.

Table 5.1 is a tool developed to prevent accidents during experiments and fieldwork in academic institutions, such as universities and research institutes, for safety professionals, faculty, staff, and students who are studying EHS. This tool is an improved version of the tool published by *Scientific Reports* in 2022.

Safety culture is defined as a culture that can be achieved by the effective systematic application of the SMS by the organization, so that all the accident-contributing factors listed in Table 5.1, which are human factors closely related to accidents, are shared by individuals and the organization, and then problems are corrected and safety is improved.

## References


Adams, M.R. (2006). *Shipboard Bridge Resource Management*. Eastport: Nor'easter Press.

Caldwell, J.A. and Caldwell, J.L. (2003). *Fatigue in Aviation: A Guide to Staying Awake at the Stick*. Surrey: Ashgate.

Endsley, M.R. (2000). Theoretical understandings of situation awareness: a critical review. In: *Situation Awareness Analysis and Measurement* (ed. M.R. Endsley and D.J. Garland), 3–32. Mahwah: Lawrence Erlbaum Associates, Inc.

Fukuoka, K. (2015). Visualization of holes and relationships between holes and latent conditions. In: *Activities in Navigation: Marine Navigation and Safety of Sea Transportation* (ed. A. Weintrit), 215–221. London: CRC Press Taylor & Francis Group.

Fukuoka, K. (2019). *Safer Seas: Systematic Accident Prevention*. Boca Raton: CRC Press Taylor & Francis Group.

Fukuoka, K. and Furusho, M. (2022). A new approach for explosion accident prevention in chemical research laboratories at universities. *Scientific Reports* 12 (1): 3185.

Grech, M.R., Horberry, T.J., and Koester, T. (2008). *Human Factors in the Maritime Domain*. Boca Raton: CRC Press Taylor & Francis Group.

Hawkins, F.H. (1987). *Human Factors in Flight*. Aldershot: Gower Technical Press.

International Civil Aviation Organization (2001). Manual of Aircraft Accident and Incident Investigation: Part 1 Organization and Planning.
International Maritime Organization (2000). Guidelines on Ergonomic Criteria for Bridge Equipment and Layout. MSC/Circ.982.
International Maritime Organization (2001). Guidance on Fatigue Mitigation and Management, MSC/Circ.1014.
International Maritime Organization (2003). Role of the human element, MSC 77/17.
ISO/IEC Guide 51: 1999(E) (1999). *Safety Aspects-Guidelines for their Inclusion in Standards*. Geneva: International Organization of Standardization and International Electrotechnical Commission.
Japan Transport Safety Board (2014). Marine accident investigation report: collision between container vessel *KOTA DUTA* and cargo vessel *TANYA KARPINSKAYA*, MA2014-5.
Lippmann, J., Taylor D, McD., Stevenson, C. et al. (2016). The demographics and diving behaviour of DAN Asia-Pacific members with and without pre-existing medical conditions. *Diving and Hyperbaric Medicine* 46 (4): 200–206.
Moore-Ede, M. (1993). *The 24 Hour Society: The Risks, Costs and Challenges of a World that Never Stops*. London: Judy Piatkus Ltd.
National Academy of Sciences (2011). *Prudent Practices in the Laboratory: Handling and Management of Chemical Hazards*. Washington, DC: The National Academies Press.
National Transportation Safety Board (1990). Marine Accident Report: Grounding of the U.S. Tankship *EXXON VALDEZ* on Bligh Reef, Prince William Sound near Valdez, Alaska. March 24, 1989. Report No NTSB/MAR-90/04.
Nishino, S. (2017). *The Stanford Method for Ultimate Sound Sleep*. Tokyo: Sunmark Publishing. (Japanese).
Occupational Safety and Health Administration, USA (2013). Appendix A to §1910.1450-National Research Council Recommendations Concerning Chemical Hygiene in Laboratories (Non-Mandatory).
Parrott, D.S. (2011). *Bridge Resource Management for Small Ships: The Watchkeeper's Manual for Limited-Tonnage Vessels*. Blacklick: The McGraw-Hill Companies.
Reason, J. (1997). *Managing the Risks of Organizational Accidents*. Farnham: Ashgate Publishing Limited.
Reason, J. (2008). *The Human Contribution: Unsafe Acts, Accidents and Heroic Recoveries*. Surrey: Ashgate Publishing Limited.
Swift, A.J. (2000). *Bridge Team Management: A Practical Guide*. London: The Nautical Institute.
UC Center for Laboratory Safety (2016). Report to the University of Hawaii at Manoa on the Hydrogen/Oxygen Explosion of March 16, 2016: Report 2: Recommendations for Improvements in UH Laboratory Safety Programs.
US Chemical Safety Board (2010). Texas Tech University Laboratory Explosion, No. 2010-05-l-TX. https://www.csb.gov/texas-tech-university-chemistry-lab-explosion/ (accessed 10 May 2023).
US Chemical Safety Board (2021). Chemical Reaction, Hydrogen Release, Explosion, and Fire at AB Specialty Silicones. https://www.csb.gov/ab-specialty-silicones-llc/ (accessed 10 May 2023).
Yamada, S. (2006). *Introduction of TQM Quality Management*. Tokyo: Nikkei Co Limited. (Japanese).
Yokomizo, Y. and Komatsubara, A. (2013). *Ergonomics for Engineers*. Tokyo: Japan Publication Service. (Japanese).

# 6

# Accident Investigation and Analysis

## 6.1 Introduction

The accident investigation process consists of data collection, analysis, and report writing. Inadequate data collection can affect the quality of the investigation report. According to the International Civil Aviation Organization (ICAO 2001), the outcome of an accident investigation depends on the skills and experience of the investigators assigned to it. In addition to technical skills and experience, accident investigators need integrity and impartiality regarding the data collected, logical thinking and patience in the search for causes and contributing factors, and the ability to establish good relationships with those involved in the accident.

This chapter describes the accident investigation process for serious accidents, such as fires, explosions, fatalities, and serious injuries, and near misses that may have developed to these serious accidents. It covers the safety of accident investigation personnel at the accident scene, the preservation and collection of evidence, the evaluation of evidence, methods of analysis, and the preparation of accident investigation reports. A simplified approach, as described in this chapter, can be used in cases of nonserious accidents.

## 6.2 Points to Be Considered When Conducting Accident Investigations at Universities, etc.

When investigating accidents that occurred at universities, etc., the scientific approach of the investigation is basically the same as that used in other industries, particularly in the aircraft and marine sectors, but it should be borne in mind that there are peculiarities arising from the purpose for which universities, etc. are established. In laboratories at universities and other institutions, researchers are working on cutting-edge technologies that are not available to the public. When preparing accident investigation reports, it is necessary to focus only on the processes related to safety without disclosing such state-of-the-art technology. To ensure that personal information is not disclosed, details should be provided anonymously rather than by individual name. Common nouns, such as university, college, faculty, researcher, postgraduate, or student, should be used instead of personal names.

In addition to these particularities, for safety investigations to be widely accepted, universities, etc. and accident investigation personnel have to implement the rules agreed in the field of aircraft and marine accident investigation to distinguish them from criminal investigations. At the beginning of an accident investigation, e.g. in the field, the accident investigation personnel must make

*Accident Prevention and Investigation: A Systematic Guide for Professionals, Educators, Researchers, and Students*, First Edition. Koji Fukuoka.

it clear to the persons involved in the accident who are to be interviewed what the purpose of the accident investigation is. In other words, they must distinguish it from a criminal investigation by explaining that the purpose of the safety investigation is not to punish people, but to prevent the accident from happening again. In addition, accident investigation reports prepared by universities, etc. need to state it clearly.

Second, accident investigation analysis needs to be standardized. In order for universities, etc. to collect reliable accident data in the future, standards need to be set so that universities, etc. can come together and standardize the accident models used as theory, the approach to accident investigation analysis, and the content that should be included in accident investigation reports.

As stated by the ICAO, the quality of accident investigation reports depends on the skills of the accident investigators who investigate and analyze accidents. Therefore, having established the above criteria, it is necessary to provide regular education and training for accident investigation personnel at universities, etc. as a way of sharing information on accident investigation.

## 6.3 Accident Investigation at the Accident Site

### 6.3.1 General Health and Safety Considerations

The scene of an accident investigation is the immediate aftermath of an accident with risks of various types and magnitudes. As stated in the safety management system (SMS) in chapter, safety is the responsibility of each individual, but organizations or institutions that direct accident investigation personnel have a responsibility to establish procedures for effective and safe accident investigations and to provide appropriate education and training for accident investigation personnel.

Accident investigation personnel should assess the risks they face before entering the site. Ongoing risk assessment will ensure the health and safety of themselves and the investigation team. The principles of site safety are as follows (Fukuoka 2019).

1. Identify all foreseeable risk factors.
2. Unacceptable risks are completely eliminated.
3. The risk assessment process reduces the risk to an acceptable or ALARP (as low as reasonably practicable) level.

Before starting an investigation at the site of the accident, the accident investigation personnel have to carry out a general or static risk assessment. Then, during the on-site investigation, the accident investigation personnel need to carry out a dynamic risk assessment at the accident site.

The static risk assessment is based on the information available before starting the accident investigation on-site. It includes the contents of the preliminary report notifying the accident, information from occupational physicians, facility managers, faculty, and students at the universities, lists of chemicals and other stored materials, the location of the accident, and expert advice on hazardous materials.

Dynamic risk assessments are based on hazards identified at the scene, detailed information from police and fire services, medical information from doctors and medical staff, and information from other professionals on-site. In the case of serious accidents involving fire, explosion, or drowning at sea, rescue operations and accident investigations are carried out by the administrative bodies, which are the fire service, police, coast guard, and accident investigation board. The investigating authorities conduct accident investigations only after rescue operations have been completed and the risk of fire, explosion, etc. have been sufficiently reduced. Therefore, when universities, etc.

conduct their own safety investigations to fully investigate the factors contributing to the accident, it is advisable to seek information about the risks from the investigating authorities, even if the risks are within acceptable levels.

## 6.3.2 Hazards to Consider

### 6.3.2.1 Environmental Hazards

#### *6.3.2.1.1 Location*

If an accident occurs during field research activities, the nature of the hazard will differ depending on whether the accident site is in the sea, a lake or swamp, or in the mountains. In the sea, investigations at the scene of an accident during periods of strong tides and currents can be a hazard for missing, and in weather conditions with high waves and strong winds, there is a risk of being involved in a secondary disaster of drowning.

On mountain expeditions, safety is largely dependent on the local weather. It is important to check the local weather station for avalanche warnings, heavy rain, fog and snow warnings, and approaching typhoons and tropical cyclones. Failure to do so may result in the risk of being caught in an avalanche or being caught in a rain or snowstorm, which can cause physical exhaustion and weakness, hypothermia, and freezing to death. Even fieldwork on level ground is subject to hazards, such as drowning in floods and river flooding caused by localized torrential rainfall. For these natural hazards, hazard maps are sometimes available from national meteorological services, which show flooded areas, areas at risk of landslides, flat areas where tsunami disasters may occur, and so on.

When investigating accidents in extremely cold or tropical climates, temperature, humidity, and wind speed are hazards, and there is a risk of frostbite, dehydration, or heatstroke if no measures have been taken against cold or extreme heat. If the results of the dynamic risk assessment of the accident site indicate that it is necessary to wear personal protective equipment (PPE), it is necessary to drink sufficient fluids before donning the PPE to prevent heat stroke. It is also necessary to ensure adequate rest during the investigation and to limit the time PPE is worn depending on the investigation situation.

#### *6.3.2.1.2 Security*

When conducting accident investigations in areas where civil war or conflict has occurred or is ongoing, poor security can be a danger, and armed forces may attack. The security situation in countries around the world can be checked on your country's foreign ministry website.

#### *6.3.2.1.3 Wildlife and Marine Life*

There are wild animals, reptiles, and marine life on land and in the sea that are harmful to humans and can be a hazard. Information should be obtained from the local authority, local guides, and dive shops who are familiar with the local situation, and the site of fieldwork should be examined to determine whether it is a habitat for harmful animals or other organisms.

### 6.3.2.2 Physical Hazards

#### *6.3.2.2.1 Damaged and Unstable Debris*

In the case of an explosion accident in a laboratory, the equipment and instruments are severely damaged, exposing sharp, and broken edges. Accident investigation personnel may be injured if they accidentally put their hands on damaged parts. When standing on high, damaged, and unstable debris in a laboratory, the debris may tip over and they may fall.

#### 6.3.2.3 Biological Hazards

##### *6.3.2.3.1 Blood-borne Infection*

In the field, in the laboratory, or during field research activities, in many cases, only colleagues who are working together on the research are present at the scene from immediately after an accident occurs until the rescue organization arrives. It is important to remember that blood and body fluids are a hazard when a casualty is present in the laboratory or field site immediately after an accident, particularly when blood is flowing from the human body or has been spread in the environment. In addition, there is a risk of contracting a number of infectious diseases, such as hepatitis B, hepatitis C, and HIV, if contact is made. After a casualty has been transported to the hospital, the bloodstain powder that adheres to desks, chairs, equipment, etc. and is easily dried and peeled off becomes a hazard, and there is a risk of blood-borne infection if the powder enters the body through the mouth, nose, eyes, or wounds.

Accident investigation personnel should be aware that they are always exposed to the risk of biological hazards, as it is impossible to confirm that blood and body fluids left at the scene are infected with, for example, hepatitis B. Even when conducting on-site investigations, disposing of wreckage, or examining parts of wreckage away from the accident site, it is necessary to be aware of these risks and to conduct investigations accordingly.

Areas where blood or other contaminants have been found must be roped off and only those wearing PPE should be allowed to enter. When handling wreckage in areas where there is a risk of infection, the number of accident investigation personnel wearing PPE should be kept to a minimum.

##### *6.3.2.3.2 Viruses and Infectious Diseases*

Information about viruses and infectious diseases that are prevalent in the region can be found on the websites of the health authorities in your country; the US Centers for Disease Control and Prevention and the European Centre for Disease Prevention and Control. If the accident site is abroad and you are traveling to an area where infectious diseases are prevalent, it is advisable to get vaccinated in advance.

#### 6.3.2.4 Substance Hazard

##### *6.3.2.4.1 Hazardous Chemicals*

In the event of a fire, explosion, or personal injury caused by a chemical, accident investigation personnel must obtain information on the name and hazards of the chemical from the laboratory where the accident occurred. During the on-site investigation, the hazardous chemical must already have been rendered harmless. However, when entering the laboratory, appropriate measuring equipment should be used to check that no hazardous substances remain in the sealed area or that low oxygen concentration conditions do not persist and present a hazard. Wear PPE that is appropriate to the hazard situation at the site and always carry out the investigation in well-ventilated conditions.

### 6.3.3 On-site Risk Assessment

A risk assessment is always carried out before a site survey is undertaken. Wearing PPE is a form of risk reduction, but only as a last resort after other risk reduction measures have been taken.

Although the extent to which PPE will be worn is predicted from the limited information obtained before the start of investigation, the result of the risk assessment may differ from the previously obtained hazard situation due to observations of the laboratory on-site and accident site conditions. In such cases, instead of wearing a higher level of PPE, it may be necessary to suspend the site investigation and wait until the risk can be reduced. This is because wearing a higher level of PPE may affect a person's mobility and, in hot and humid conditions, the need to manage time to prevent dehydration and heat stroke may impede the progress of the site investigation.

If new hazards are identified during a site investigation, the appropriate PPE must be worn. For example, if a large amount of blood is found on the laboratory floor during a site investigation and the investigation personnel are wearing helmets, gloves, and goggles, there is a risk of blood-borne infection, and the appropriate PPE must be worn. This means that protective clothing and masks are added as new PPE. In this case, the additional PPE is not worn by everyone, but only by a limited number of people who enter the same laboratory to investigate. Only those authorized personnel will wear the additional PPE.

### 6.3.4 Risk Assessment Sheet

If there are residual risks at the site where an accident investigation is to be initiated, such as a fire, explosion, casualty, or injury accident in a laboratory or fieldwork, a risk assessment sheet is useful as a risk assessment during the site investigation. The sheet includes risk category, hazard type and hazard situation, location of identified hazards, risk level, priority number, and risk controls.

Under the risk category, enter the name of the hazard, e.g. flammable, structural, and explosive. Under hazard type, if the category is flammable, enter the suspected hazard, e.g. fuel and fuel tanks, flammable liquids, oxygen leaks, etc. Under hazard situation and location of identified hazards, enter the unsafe condition and its location. The risk level could refer to the risk matrix presented in Chapter 4, but the risk assessment sheet must be completed in a short time. Therefore, a simplified version of the risk level can be used where the severity and frequency scales are measured on three scales and multiplied by the estimated values.

Risk reduction is carried out according to the priorities of the risk reduction methods presented in Chapter 2. For example, if there is residual fuel in the tank and it is extracted as a part of the risk control process, this is described as fuel extraction. If the fuel tank is then moved to a designated hazardous materials storage warehouse, this is described as such, and the risk level associated with the fuel is reduced.

Accident investigation personnel must follow the same process as above to continually monitor risks and ensure the safety of the accident investigation personnel conducting the site investigation and the people involved, even if new hazards are discovered during the investigation.

### 6.3.5 Cordons

Cordons are restricted areas of human movement set up primarily to preserve evidence and are essentially of two types: (1) inner cordons and (2) outer cordons. In the event of a major transport accident, an inner cordon is set up around the main damaged areas, such as the hull of a ship, aircraft fuselage, train carriages, and scattered debris. Outside of this area, an outer cordon is set up to restrict access to the media and other outsiders. Accidents that occur in universities and academic institutions do not have such a large cordon, but signs and ropes are placed in and around the damaged laboratory to clearly indicate that the area is off limits, also to prevent secondary accidents.

### 6.3.6 PPE and Blood Contamination

The ICAO (2001) recommends the following site safety precautions for contact with blood and body fluids.

- Areas contaminated with blood or body fluids should be clearly marked by ropes or signs indicating where there is only one entrance and exit.
- Only people wearing PPE correctly should enter the area; when wearing PPE, eating, drinking, touching the face, eyes, mouth, or nose is prohibited.
- It is important to control working hours when using PPE as heat stroke can occur in humid and hot weather.
- Exposed skin should be wiped with a damp towel and washed with soap and water or disinfectant.
- Appropriate PPE for biological hazards involving blood and body fluids includes disposable latex gloves worn under work gloves, face masks covering the nose and mouth, safety glasses with one-way check valves or vents, durable and liquid-resistant disposable protective clothing, disposable polyvinyl chloride or butyl rubber shoe covers, and leather or rubber safety boots. In addition, disinfectants containing 70% alcohol and disposal bags with red or orange biohazard labels are required.

In addition to these recommendations, PPE must be donned and removed correctly to protect accident investigation personnel from harm. The following are key points to consider when donning and undressing PPE.

- When wearing PPE, a buddy system should be used to ensure that PPE is worn correctly by both parties or is not damaged.
- Goggles may fog up due to air leakage between the nose and eyes. If this happens, bend the mask so that it is in close contact with the face.
- To remove protective clothing, roll the hood from the inside out, then the torso and legs from top to bottom.
- Disposable gloves are usually worn with two gloves of different colors. If one glove is damaged, the difference in color will be noticed, and the second glove will prevent blood contamination. The gloves are removed from the wrist area and wrapped from the inside out.
- If it is necessary to remove PPE, such as at the end of a site investigation or during a temporary break, it should be done in a designated area using a specified removal method. Disposable gloves and masks shall be disposed of in a designated waste bag to avoid cross-contamination.
- After disinfection, accident investigation personnel should remove their protective clothing and take a break in a designated area. Ensure that the person wearing the protective clothing does not mix with the person removing it during the break.

### 6.3.7 Evidence

#### 6.3.7.1 Evidence in General

The importance of evidence is to establish the chronological sequence in which accidents occur, to support robust analysis, and to justify recommended safety measures. Evidence falls into four categories: (1) human evidence, (2) physical evidence, (3) documentary evidence, and (4) electronic evidence.

The principle of evidence collection is to collect evidence in the order in which it is most likely to be lost. Evidence that is likely to be lost includes human evidence, damage to the scene immediately after the accident, broken metal surfaces of equipment and high-pressure gas cylinders, leaked liquids and chemicals, oxygen and certain gas concentrations in the laboratory, traces on the ground, and the weather and sea conditions at the time of the accident.

Evidence is collected using the software, hardware, environment, liveware (SHEL) model to ensure that all elements are comprehensively collected. The first step should be to establish the situation at the scene of the accident and then to interview the person or operator who caused the accident in order to obtain an early overall picture of the accident, which will speed up the subsequent collection of evidence and the identification of the causes and contributing factors of the accident.

Common evidence that should be collected includes the following:

- Human evidence by interviewing operators and persons involved in the accident.
- Photographing debris and damage at the accident site, e.g. in a laboratory.
- Drawing up a sketch map of the accident site.
- A diagram showing the arrangement of equipment and instruments in the laboratory at the time of the accident.
- Measurement of the relative positions of injured persons and persons involved in the accident and drawing of a layout map.
- Physical evidence of experimental materials left at the scene, including chemical substances.
- Experimental plans, emergency response plans, standard operating procedures (SOPs), protocols, and other rules and regulations of the laboratory and organization relating to safety management.
- Documents recording the details of the experiment being conducted at the time of the accident and the time of performance for each subject.
- Collection of documents such as records of group meetings related to the experiment, tables showing the hours worked by the person in charge of the experiment and the team (shift tables), and records of equipment maintenance and inspections.
- Operating manuals for equipment, Chemical Hygiene Plan, records of training and education, records of internal and external audits, and diagrams showing the structure of the organization.
- Electronic data on the content of experiments and the atmosphere of the laboratory as recorded on equipment, apparatus, PCs, and other electronic devices.
- Documents of approval and licensing by national organizations, including certificates of competence, required for the conduct of experiments.
- In the case of field research, weather forecasts and actual weather reports, current forecasts and actual current and tidal currents, and geographical features of mountainous and flat areas.
- Reenactment of the accident by witnesses or those conducting the experiment at the time.
- Human evidence from interviews with personnel of the organization involved in the accident and organizational factors related to the SMS.

Cross-checking evidence is important in the process of collecting evidence. Cross-checking means that the evidence collected by the accident investigation personnel contains the same information from different sources. The content of the statements made by operators and those who were with them at the accident site must match and corroborate the physical and electronic evidence. When all the evidence has been collected, the accident investigation personnel will check the evidence collected by the accident investigation team using the SHEL model and, if any evidence is missing, will collect it again. This allows them to identify necessary and missing evidence at an early stage of the accident investigation process and to collect evidence comprehensively. As the investigation progresses, they can exclude information that is irrelevant to the accident.

#### 6.3.7.2 Physical Evidence

Once physical evidence has been moved or removed from the scene of an accident, it cannot be returned exactly to its original position. The following points should be noted:

1. Any removal of physical evidence must be agreed upon in advance with the laboratory faculty and others involved.
2. Physical evidence should not be removed until interviews with operators, laboratory faculty, and lab personnel have been completed. This is because the scene of the accident may stimulate the memory of witnesses.
3. Prior to removal, the physical evidence should be measured, photographed, and its location should be recorded on a sketch of the scene.
4. Care should be taken not to damage impact marks or broken surfaces during removal. Damaged or broken evidence should be protected from further damage. For technical examination, fragile pieces of evidence should be placed in padded boxes, labeled, and sent to a laboratory for accurate analysis (International Maritime Organization [IMO] 2014a).
5. When accident investigation personnel come across damaged equipment or tools, researching and obtaining similar or the same type of replacements can help them to get a complete picture of the overall condition. Obtaining and studying system diagrams and blueprints can help to understand what the original condition should have been.

##### *6.3.7.2.1 Bodies*

In cases where a criminal investigation is being carried out by a government agency, an autopsy may be performed, depending on the circumstances. This may provide important clues as to the cause of death, such as toxic gases or lack of oxygen.

#### 6.3.7.3 Documentary Evidence

Documentary evidence provides important data and should be systematically preserved and protected in the same way as physical evidence. Documentary evidence includes procedure manuals, protocols, laboratory notebooks, laboratory equipment and dive computer readouts, paper or electronic maps and charts, licenses, education and training records, certificates, photographs, videos during experiments, guidelines and manuals relating to the accident, and so forth.

Planning for an accident investigation should include identifying the records to be collected and the accident investigation personnel responsible for their collection. Documentary evidence pertaining to safety management produced by universities, etc. organizations, and environmental, health and safety (EHS) can provide important clues as to accident-contributing factors and organizational factors.

##### *6.3.7.3.1 Environment*

For field research activities, the environment and meteorological and oceanographic conditions are important factors in the occurrence of accidents, mitigation measures, and the implementation of contingency plans. There are three types of meteorological and marine weather evidence: forecasted meteorological and oceanographic conditions, eyewitness reports of meteorological and nautical conditions, and actual meteorological and marine weather conditions.

It is advisable to obtain meteorological and oceanographic information from as many sources as possible, as the information obtained from interviews may not be consistent with other sources. In the event of a person missing at sea or falling from a vessel, seawater temperature is the most important factor in determining survival. Detailed and accurate data on currents, tides, wave heights, wind strength, and direction can be obtained from meteorological services, coast guards, etc.

#### 6.3.7.3.2 Photography

Photography serves as a permanent visual record of the accident scene as well as evidence and verification. When photographing the scene of an accident in a laboratory, it is necessary to first take photographs from a distance to capture the entire laboratory, and then move closer to capture the overall view of the equipment involved in the accident, and finally move closer to capture the situation at the point where the equipment has been damaged. Generally, the scale is used in the following order: overview from a distance, then medium range (3–6 m from the subject), and then close up (1.5 m or less; IMO 2014a).

Accident investigation personnel should be aware that in some universities, etc., closed circuit televisions or surveillance cameras are installed at building entrances and corridors to record the behavior and situation before and after an accident. These images may record the circumstances in which the accident develops, as well as firefighting and evacuation activities.

In accidents involving a diving boat or a vessel owned by universities, etc. where the vessel has sunk, preaccident photographs and data of the vessel are required to analyze the stability of the vessel prior to the accident. By examining and analyzing the photographs and data, changes to the hull, structure, and equipment can be identified. The accident investigation agency may also use remotely operated vehicles or underwater drones to determine the location of the sinking and the condition of the vessel.

#### 6.3.7.4 Electronic Evidence

Electronic data is evidence that is easily lost and should be reproduced or downloaded as soon as possible after an accident. Because of the special nature of laboratory equipment and the devices handled in the laboratory, permission to collect electronic data should be obtained from the relevant laboratory faculty member and a request to retrieve the data should be made to the person in charge in the laboratory.

## 6.4 Witnesses Interviewing

### 6.4.1 Human Evidence

Human evidence is the most volatile form of evidence, and interviews are a means of obtaining human evidence. Interviewing is a difficult skill that requires skill and practice, and PEACE (acronym for Prepare and plan, Engage and explain, Account, Clarification and challenge, Closure, and Evaluation) is a cognitive interviewing technique that can elicit a great deal of factual information from the interviewee in a short time (Fisher and Geiselman 1992). This section introduces useful parts of PEACE that can be used for accidents at universities, etc. This section describes how to collect human evidence from people involved in the accident, including the operator who caused the accident, other witnesses during the experiment, faculty and PIs in the laboratory, safety managers, top management of the organization, and other parties involved in the accident.

#### 6.4.1.1 Prepare and Plan

Before the on-site investigation of the accident, it is useful to identify the interviewee and develop a brief plan covering the topic and areas of questioning. The approach to the interview will depend on the interviewee. Those directly involved in the accident are key witnesses and may be asked detailed questions about the event and the human factors that contributed to the accident.

Prior to the interview, it is important to collect, analyze, and evaluate available evidence, such as obtained experimental plans, meteorological, and oceanographic data, etc. Researching and analyzing whether similar accidents have occurred in the past can help accident investigation personnel prepare to conduct a site investigation and develop safety recommendations.

Milne and Bull (1999) state that when interviewing witnesses, it is important to manage the conversation during the interview, and the key points are the following: (1) The interview is a conversation with a purpose. (2) The interviewer needs to have some control. (3) To be in control, the interviewer must prepare and plan the interview carefully. (4) Preparation includes reading, analyzing, and understanding existing evidence, making a list of key points to be proved, and preparing a framework of general questions.

#### 6.4.1.2 Engage and Explain

This phase marks the beginning of the interview. The accident investigation personnel introduce themselves, including the members of the accident investigation team, and greet the interviewee by name. Starting with neutral questions unrelated to the accident will help to ease the interviewee's nervousness and allow both parties to interact as naturally as possible. Four issues should be clarified at the start of the interview.

1. Powers and duties of accident investigation personnel
2. Purpose of the interview
3. Handling of personal information
4. Results of the information obtained during the interview

The gist of what to say to the interviewee on the basis of the above is as follows:

"In accordance with the health and safety committee regulations, we, the accident investigation team, will be investigating the accident that occurred at (location of accident site) this time. The purpose of the accident investigation is to prevent a recurrence, not to punish people. The results of the accident investigation will be compiled into an accident investigation report, in which personal information, such as names of individuals and laboratories, will be kept anonymous and will be made public at the health and safety committee."

If a digital recorder is to be used in an interview, the accident investigation personnel must obtain the consent of the interviewee. Digital recorders allow the interviewer to concentrate on the conversation with the interviewee and the interview plan. However, digital recorders also have disadvantages. The interviewer may become too dependent on the recorder, and conversations may not be recorded if the recorder is accidentally placed under a document.

If a digital recorder is used, the interviewer must record the following items before and after the interview. Before the interview, record the date and time of the interview, the location where the interview takes place, the name of the interviewer, the name of the interviewee, and the names of any other persons present at the interview. After the interview, record the time at which the interview ended.

#### 6.4.1.3 Account, Clarification, and Challenge

This stage is the main part of the interview, and it is useful to utilize cognitive interviewing techniques. Some useful cognitive interviewing techniques are as follows (Fisher and Geiselman 1992).

1. Talk to the interviewee throughout the interview to actively narrate factual information, rather than waiting for the interviewer to ask questions before starting to talk.
2. Ask open questions. Open questions are the best form of questioning for gathering evidence; they avoid the risk of imposing the interviewer's point of view. The following are examples of open questions. "Tell me everything," "Please explain in detail," "Please explain in as much detail as possible."

   Closed questions are the second-best question format because they are less informative than open questions and can lead to inaccurate answers. However, the use of open questions alone will not complete the interview, and these two question formats need to be combined appropriately to obtain more accurate and detailed human evidence.
3. Listen actively.
4. Do not interrupt. However, if the content of the conversation is far from the purpose of the interview, the interviewee can be interrupted in order to control the interview and direct the conversation in the correct direction for the purpose.

Cognitive interviewing techniques cannot be used with uncooperative people or those with language barriers. The interviewer can assist in the gathering of evidence by showing the interviewee written records or other material relevant to the accident, or by having the interviewee draw a picture. To take the example of an investigation into a serious near miss, if an abnormal explosion sound occurred during an experiment using chemical substances, the interviewer can help the interviewee retrieve a more accurate and detailed memory by moving the interview to the location where the unusual explosion sound occurred and asking the questions. The interviewee may be able to recall the circumstances of the relevant topic, such as the procedures and conditions under which the chemical was handled at the time of the near miss, the quantities used, the equipment used, and so on. This method, which is conducted on-site, helps the interviewee to recall accurate and detailed memories of the decision-making process at the time of the event.

##### *6.4.1.3.1 The Subjects to Be Interviewed*

First, the history of the events leading up to the accident is interviewed to get an overall picture of the accident. The history of the accident includes the purpose of the experiment and its methods, such as procedures, the state of preparation before the start of the experiment, the process by which the accident occurred, and the emergency response when the accident occurred. Each of the following elements of the SHEL model is then interviewed in detail. These list some, but not all, of the items that are accident-contributing factors. It is advisable to modify or add to the content according to the type of accident.

##### *6.4.1.3.2 Factors of an Operator*

- Hours of work or study and rest in the last few days.
- Sleepiness and fatigue.
- Medical fitness, drug and alcohol use, especially for field research activities, and scuba diving.
- Psychological factors such as time pressure.
- Psychological factors such as assumptions, complacency, and distraction.
- Omission of processes due to forgetfulness or illusion.

- Assigned tasks and actual behavior at the time of the accident.
- Workload and task complexity.
- Level of situational awareness of possible hazards in the experiment (see Endsley's situational awareness diagram in Figure 5.1 in Chapter 5).
- Risk assessment, including risk reduction methods implemented in the experiment.
- Competence, skills, knowledge, and experience.
- Content and attendance of education and practical training for the experiment relevant to the accident.

#### 6.4.1.3.3 Laboratory Safety Management

- Laboratory management and supervision, allocation of duties and responsibilities.
- Composition of research team members and status of teamwork.
- Status of development and implementation of experimental plans, procedures, and specific SOPs for the relevant experiments.
- Communication among research team members or between team members, faculty, and PIs.
- Provision of opportunities for safety-related communication.
- Status of education and practical training in the laboratory for experiments related to the accident.
- Whether or not any near misses have occurred in the laboratory and, if they have, what action has been taken and what measures have been implemented.
- Development of emergency response plans and training; status of implementation at the time of the accident.

#### 6.4.1.3.4 Factors of Hardware

- Status of equipment design, including inherently safe design, fool proofing, and fail-safes.
- For self-built laboratory equipment, safety standards and risk reduction measures for the equipment.
- Status of maintenance and inspections, including mandatory and voluntary inspections.
- Status of design modifications and repairs.
- Use equipment that is compatible with the substances used in the experiment.

#### 6.4.1.3.5 Working Conditions

- Ergonomic elements of the laboratory or equipment relevant to the accident.
- Vibration, noise, temperature, humidity, and lighting conditions in the laboratory.
- Atmospheric conditions in the laboratory, such as oxygen concentration, toxic gas concentration, and lower explosive limits.
- Ventilation conditions in the laboratory.

#### 6.4.1.3.6 Environmental Factors

- Weather and sea conditions.
- Submarine topography, special topography of mountainous areas, such as topographic features causing strong currents, avalanches, landslides, torrential rainfall causing floods, and river flooding and landslides.
- Whether any accidents have occurred in the same land or marine area and the factors that led to the accident.

#### 6.4.1.3.7 *Safety Management in the Organization*

- Status of established criteria for suspension of experiments.
- Status of development and implementation of general and specific SOPs, monitoring, and review systems.
- Procedures for modification of experimental plans.
- Status of statutory maintenance and inspection of facilities and equipment.
- Establishment and provision of customized education and training programs required for the laboratory, and the status of attendance and follow-up of these programs.
- Specific activities of the health and safety committee in relation to the experiment; approval of the experimental plan, including any amendments.
- Status of workplace inspections, internal audits, and their follow-up.
- The system for the conduct of accident investigations in the event of accidents or near misses and the status of the application of lessons learned from accidents.
- Formulation of emergency response plans and operational structures.
- Powers and duties of the safety management office (EHS) and the status of their implementation.
- Specific EHS activities related to this experiment.
- Document management system to avoid mixing old SOPs with new ones.
- Role and activities of the risk management committee in relation to prevention of accident recurrence.
- Communication between top management, EHS, and laboratories.
- Top management policy and commitment to EHS.
- EHS implementation and organizational structure.

#### 6.4.1.3.8 *External Influences*

- Existence of rules of international organizations, rules and regulations of national regulators and restrictions, bylaws of local authorities, guidelines of private and academic organizations, and their relevance to this accident.

### 6.4.1.4 Closure

At the end of the interview, the interviewer can thank the interviewee for the evidence they have provided and remind them that they are an important part of the investigation.

The interviewer can encourage the interviewee to report back to the interviewer any information that comes to mind after the interview.

### 6.4.1.5 Evaluate

Examine the content of the interview and decide whether further investigation is needed. The content of the interview should be compared with other evidence.

#### 6.4.1.5.1 *Use of an Interpreter*

As a general rule, interviews should be conducted in the interviewee's native language, but if the interviewee is a nonnative speaker of English and has limited ability to speak English competently, it is necessary to have a native language interpreter. National accident investigation agencies often employ interpreters to conduct interviews, but the following points should be observed when using interpreters, even at universities, etc.

1. Provide a list of technical terms relevant to the accident for the interpreter to prepare.
2. Allow the interpreter at least twice the usual time for the interview.
3. Explain interview techniques to the interpreter.
4. Ask the interviewee to speak in short segments so that the interpreter can interpret.
5. Instruct the interviewee to speak in short segments so that the interpreter can interpret.
6. The interpreter is instructed to translate everything the interviewee says.
7. The interviewer should speak directly to the interviewee, not through the interpreter.
8. The interviewer should ask the interpreter for his or her impressions of the interviewee after the interview. This will help the interviewer to determine whether the interviewee is dictating factual information.

## 6.5 Assessing the Evidence

Before analysis, the evidence collected should be assessed. Electronic evidence, such as equipment and instrument readings, is generally objective, accurate, and reliable, making a significant contribution to the identifying of accident-contributing factors. However, accident investigation personnel need to understand the limitations of this evidence. Some equipment records data intermittently rather than continuously and displays it on a screen. If equipment is not properly calibrated, the measurements themselves may be erroneous.

Human evidence from interviewees depends on the number of days elapsed between the date of the accident and the date of the interview and the skill of the accident investigation personnel conducting the interview. Therefore, the value of human evidence is lower than that of electronic evidence.

Strauch (2004) identifies five factors that affect the value of documents collected by accident investigation personnel: (1) Quantity, (2) Frequency and regularity, (3) Temporal closeness, (4) Reliability, and (5) Validity.

### 6.5.1 Quantity

For a particular piece of information, the more evidence an accident investigation personnel has, the greater the value of that evidence. In other words, if information from different types of evidence is consistent, the information is valuable and contributes to the analysis.

If electronic data, such as equipment records, are not available and only the oral testimony of one researcher is obtained, the content of that testimony is of little evidentiary value. However, if the same oral testimony is obtained from several researchers who were working as research teams in the vicinity of the laboratory at the time of the accident, the information is valuable evidence. For example, in the case of a laboratory fire, one researcher A said, "I threw a water-repellent substance into a waste bin. A short time later, smoke came out of the bin." If two other researchers give the same account as researcher A, for example, that researcher A threw away a water-repellent substance and smoke was produced, the content of the report becomes reliable.

### 6.5.2 Frequency and Regularity

If the condition is measured frequently and regularly, accident investigation personnel can identify precise trends of change. Measurements should be taken as frequently and regularly as possible.

In laboratories where liquid nitrogen or liquid helium is handled, oxygen meters are installed in the laboratory, and indicators displaying the measured values are installed inside and outside the laboratory to help prevent oxygen deficiency accidents. The instruments, when properly calibrated, allow the researcher to determine the exact change in the current oxygen concentration in the laboratory. This not only helps the researchers but also accident investigation personnel assess risks before starting work, and the measurements are treated as reliable in the event of an accident.

### 6.5.3 Temporal Closeness

The more data that is collected immediately before or immediately after an accident, the closer the value of that data will be to its actual value at the time of the accident. For example, if records of education and training in risk assessment of energetic substances are kept in a laboratory and are available immediately before an accident, they are valuable in assessing the effectiveness of that training.

### 6.5.4 Reliability

Reliability indicates that measurements are consistent and reproducible. A reliable device or measurement is reproducible, and the results will be the same no matter how many times the measurement is taken, whereas unreliability can lead to large variations. For example, weight measured on a calibrated scale is reliable because it is constant, except for the daily variations of the subject. Hearsay evidence, where a person who has not actually witnessed the accident gives indirect information about what he or she has heard from others, or oral evidence based on speculation without having seen the facts, is inconsistent and unreliable.

### 6.5.5 Validity

Suppose a student who has studied risk assessment methods in video is involved in an accident while carrying out a risk assessment in an experiment. When the accident investigation personnel investigate the factors that contributed to the accident, they will review the content of the video that the student has learned together, with the expert to assess the validity of the video. The following example is also about validity. When measuring the specific skills of researchers, licenses and certificates issued by national organizations are generally highly valid, but in some cases, certificates issued independently by a small number of private organizations are not.

## 6.6 Analysis Process

### 6.6.1 Investigation and Analysis Using the RMQMP Model

The Risk Management and Quality Management Process approach (RMQMP) model was originally developed by the author in the investigation and analysis of marine accidents (Fukuoka 2016; Fukuoka and Furusho 2016a,b; Fukuoka and Furusho 2017).

The procedures for accident investigation and analysis using the RMQMP model are detailed in the RMQMP model in Chapter 3, and the procedures for developing safety recommendations are detailed in Table 5.1 in Chapter 5.

The features of the model are as follows:

1. It allows for a comprehensive collection of evidence at the accident investigation stage.
2. The model views accidents as events that occur within the components of society and identifies which of the risk management processes and PDCA cycles used in local workplaces and organizations in laboratories and field research activities were problematic. The problems correspond to the hole of the Swiss cheese model.
3. The accident-contributing factors, which consist of the elements of the SHEL model, are provided in tabular form so that all accident-contributing factors are comprehensively captured in the on-site investigation and analysis.
4. It is also easy to understand which subjects from the accident-contributing factors in the table should be considered as safety recommendations. Potential accident-contributing factors that existed in the laboratory and organizational factors, such as universities, are covered by the safety recommendations.
5. Accidents involving more than one organization can be drawn.
6. Accident-contributing factors can be statistically processed, and the relationship between accident occurrence and accident-contributing factors can be shown by regression analysis.

### 6.6.2 Investigation Analysis Adopted by the IMO

The IMO investigation and analysis method is based on the Swiss cheese model and SHEL model presented in Chapter 1 and uses event and contributory factors charts (ECFC), which can simultaneously display time series of casualties and casualty factors. The following section describes the terminology used in the ECFC and then describes the analysis methodology.

#### 6.6.2.1 Terms and Concepts

"Event" and "condition" are interrelated: a particular condition causes an event, which leads to a new condition, and the transition is repeated.

The IMO (2014b) defines the terms used in the analysis as follows:

- Event: an action, omission, or other happening. Events are generally based on solid facts.
- Accident event: an event that is assessed to be inappropriate and significant in the sequence of events that lead to the incident and accident.
- Casualty event: the accident and incident, or one of a number of connected accidents and/or incidents forming the overall occurrence.
- Contributing factor: a condition that may have contributed to an accident event or worsened its consequence.
- Safety issue: an issue that includes one or more contributing factors and/or other unsafe conditions.
- Safety deficiency: a safety issue with risks for which existing defenses aimed at preventing an accident event, and/or those aimed at eliminating or reducing its consequences are assessed to be either inadequate or missing.

To clearly illustrate the concepts of these terms, Figure 6.1 (Fukuoka 2019) compares the terms defined by the IMO with the "Concepts of hazard and accident" and accident factors as defined in ISO/IEC Guide 51(2009).

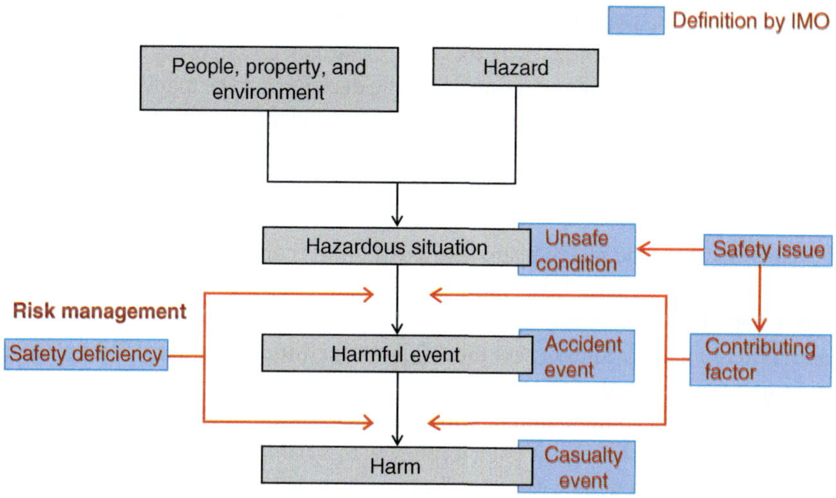


**Figure 6.1** Concept of hazard and accident and accident-contributing factors. *Source:* Adapted from Fukuoka (2019). *Note:* This figure contrasts Figure 2.1 "Concept of hazard and accident" with the terminology used in the "Event and Contributory Factors Charts," the accident analysis tool adopted by the IMO.

Harmful event is equivalent to accident event; harm is synonymous with casualty event; unsafe condition implies a hazardous situation; contributing factors are the hazardous contributing factors refer to the progression from a hazardous situation to a harmful event and from a harmful event to harm. Safety deficiencies refer to deficiencies in the defenses, such as risk management, built into the system to prevent the progression of harmful events and harm.

#### 6.6.2.2 Analysis Steps

The IMO has adopted the concept of the Swiss cheese model, but has included technical failures in the analysis process, noting that the Swiss cheese model does not place much emphasis on technical failures that are not due to operator error. The following are five-step procedures for collecting evidence and analyzing the evidence collected (IMO 2014a).

1. Establish the chain of events.

   During the on-site investigation, evidence is collected from interviews with operators, colleagues, and others involved in the accident, along with accident reconstruction, electronic evidence, damage to facilities and equipment, documents related to the accident, and a timeline is prepared. The timeline describes the events and conditions.

2. Identification of the accident event and casualties.

   Taking the example of the explosion accident at University A, the recommended dosage of nickel hydrazine perchlorate (NHP) was 100 mg. The accident event was that the NHP was used in excess of this dosage and the NHP was stirred with a pestle. The casualty event was that the NHP was stirred up with a pestle, resulting in an explosion that injured a postgraduate student.

   The criterion for deciding whether an event is an accident event is whether a similar event would be caused by a well and safely managed laboratory in another organization under similar circumstances. If the answer is no, then the event is an accident event. The decision of the accident event described above is not only carried out by the accident investigation teams but also in consultation with chemical experts and professors who can provide the material for the decision. Note that the IMO manual states that the decision is to be made by the national accident investigator, the reason being that the marine accident investigator is an expert in the maritime industry and can therefore make the decision.

3. Analysis of each accident event to identify individual behaviors and mechanisms of human error and technical failure.

   This stage identifies the mechanisms and error types that best represent human error and technical failure; the Generic Error Modelling System (GEMS) developed by Reason can be used (see Chapter 3).

4. For each individual behavioral or technical failure, the influencing local workplace environmental factors are analyzed.

   This stage involves a combination of why–why analysis and the SHEL model to investigate and analyze the potential accident-contributing factors behind each accident event in laboratory and field research activities. Hypothetical factors may be present at this stage of analysis, and further evidence is required to support them.

5. Carry out a more detailed analysis to identify the organizational factors underlying the event.

   As in step 4, use the why–why analysis and the SHEL model to identify the contributing factors within the organization. These contributing factors are the safety issues and the safety deficiencies that form the basis for the recommended safety actions. If there is no organization capable of resolving the contributing factors identified by the analysis, the accident investigation shall be terminated in accordance with the stop rule.

   The IMO recommends the use of ECFCs to provide a visual representation of the above procedure and the results of the analysis. Figure 6.2 shows an illustration of an explosion accident at University A that was investigated, analyzed, and published by the US Chemical Safety Board based on ECFCs.

#### 6.6.2.3 Characteristics of Event and Contributing Factor Charts

The ECFC (DOE 1999) helps to identify the necessary and sufficient triggering conditions and events for an accident to occur and to graphically represent the contributing factors. ECFCs provide a chronological view of the accident process and are primarily used to organize evidence of the factors underlying accident events. The advantages of this method of analysis include the following:

1. It shows the sequence of events leading up to an accident and the contributing factors.
2. It provides a communication tool between accident investigation personnel by identifying gaps in information for data collection and analysis.
3. It provides a structured method for integrating the evidence collected.
4. It presents information about the accident to guide the preparation of accident investigation reports.

On the other hand, it has the following disadvantages:

1. It cannot be applied to systemic accidents. Even where applicable, the graphical representation of events and contributing factors may be complex and difficult to understand.
2. It may be difficult to represent accidents involving multiple organizations.
3. The graphical representation relies on relative and subjective definitions of accident events as inappropriate and significant.

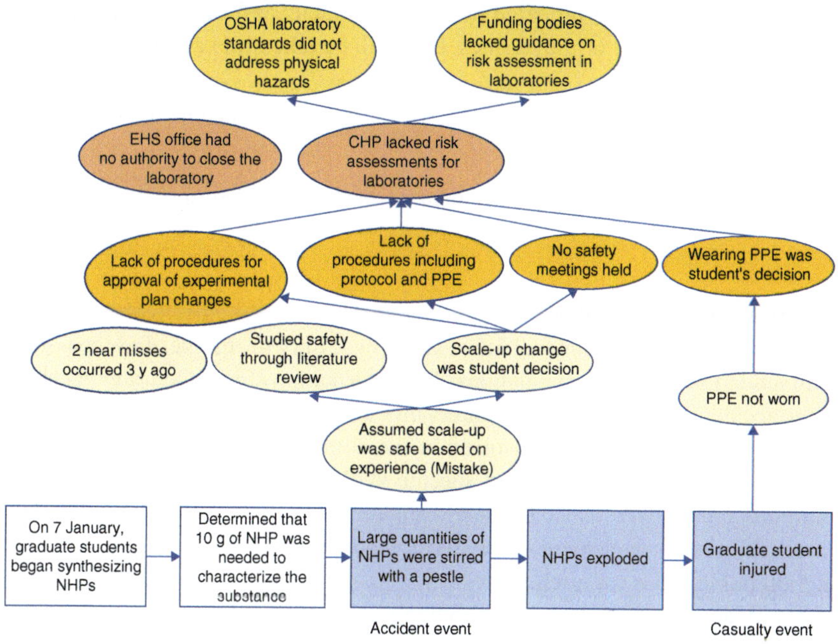


**Figure 6.2** Analysis of explosion accidents at University A using event and contributory factors charts. *Note:* The bottom layer of the diagram shows the occurrence of events in chronological order. The factors behind the occurrence of the Accident event and the Casualty event are represented hierarchically in the upper layers. The top layer represents the problems faced by external organizations in relation to this accident. Between the lowest and highest layers are the problems of the laboratory and the university.

## 6.7 Writing the Accident Investigation Report

The first thing to remember is what is emphasized in Section 6.2. The person writing the accident investigation report shall bear in mind that the purpose of writing the accident investigation report is to prevent the recurrence of the accident or incident and to mitigate the damage. In the case of organizational accidents at universities, etc., analysis of published full versions of accident investigation reports has shown that accidents are not caused by a single cause but by a combination of many contributing factors (Fukuoka and Furusho 2022).

The accident investigation report visualizes the accident-contributing factors identified in the investigation analysis and makes safety recommendations to the organization involved in the accident that can eliminate or control these accident-contributing factors. In order for the contents of the accident investigation report to be accepted by the recipients of the safety recommendations and by the faculty, staff, and students of universities, etc., it must be described in a logical manner, stating that the information contained is complete and accurate, that this information is based on evidence, and that the safety recommendations are logically derived from the investigation and analysis.

In the field of aircraft and marine accidents, the reports published by national accident investigation organizations are based on the same format worldwide, with the order of the content being standardized according to the ICAO Annex 13 to the Convention on International Civil Aviation (ICAO, 2016) and the IMO Casualty Investigation Code (IMO 2014b). The contents include the overview of the accident, Section 1: factual information, Section 2: analysis, Section 3: conclusions and causes, and Section 4: safety recommendations. The subjects to be included in each section of the accident investigation report prepared by universities, etc. and the points to be noted are as follows, and the particularities of universities, etc. shall be taken into account with regard to the content of these subjects.

### 6.7.1 Overview of the Accident

Outlines the main facts and causes of the accident or incident and describes the human and material damage caused by it.

#### 6.7.1.1 Section 1 Factual Information

The factual information shall include all information necessary to analyze, draw conclusions, and make recommended safety recommendations. The information should describe in chronological order how the accident occurred, based on interviews and evidence. It should describe the roles of the researchers, students, and others involved in the accident, their qualifications if the experiments or field research activities required specific skills, and the facts and background of the unsafe conditions and events revealed by the investigation so that the reader can understand them. It is recommended that each element of the SHEL model is included in the report to ensure that the information published is comprehensive.

Supplementary documentation such as photographs, electronic data records, and data derived from scientific experiments and studies and their conclusions are included in the text or attached as an appendix. Measures taken by the organization that caused the accident after the accident to prevent recurrence, laws related to the accident, and extracts from the organization's rules and regulations are also included in Section 1. Specifically, the following are included as follows:

1. Detailed description based on factual information, the background and circumstances of the accident or incident obtained from interviews and other sources.

   It should include the status of risk assessment, emergency treatment, and damage to persons and facilities or equipment.
2. Information and conditions of the operator.

   This includes physiological factors, psychological factors, experience, knowledge, and skills in the particular experiment, the status of implementation and participation in education and training on the general and specific SOPs.
3. Characteristics of the substances, including the chemicals involved in the accident, and information on the characteristics of the experimental sea area or region if field research is conducted.
4. The condition of facilities, equipment, and apparatus relevant to the accident.

   This includes the state of maintenance and inspection of facilities, equipment, etc., the state of standards to ensure safety in the case of self-made equipment, and the ergonomic design of facilities and equipment.
5. Information on the laboratory to which the operator belongs.

   The composition and roles of the research team, communication with the laboratory staff and between the team, the system for dealing with experiment-specific risks and the status of the development of emergency response plans and training, etc.
6. Information on universities, etc.

This includes the provision and follow-up of training programs, workplace inspections and internal audits, as well as the accident investigation system and the use of lessons learned from accidents and near misses.

7. Information and legislation of nonuniversity organizations involved in the accident, e.g. funding bodies, national health and safety authorities.
8. Where appropriate, photographs and a schematic diagram of the laboratory or site where the accident occurred, and photographs of the facilities and equipment related to the accident.
9. A brief description of similar accidents that have occurred in the past.

#### 6.7.1.2 Section 2 Analysis

The analysis provides a logical link between the factual information and the conclusions. It is not possible to include new conditions or events in the analysis topic that are not described in Section 1 factual information. The analysis identifies why the condition or event developed into an accident, why the mitigation measures did not work, etc.

Aircraft and marine accident investigation reports are similar to scientific papers in that they are logically developed on the basis of factual information. However, in the case of the former, there is often a lack of factual information due to the time constraints of producing the report, the limited scope of the accident investigation and budgetary constraints. These gaps can be filled by reasoning from the available information, using logical thinking based on the findings of academic institutions. Hypotheses must be formed through reasoning and then discussed and tested based on these hypotheses. Reasoning and unknowns should be clearly explained in the report. Accident investigation reports from universities, etc. should also be prepared on the basis of the above. The subjects of analysis are the following:

1. Analysis of operator error derived from factual information.
2. Analysis of accident-contributing factors identified at the local workplace, i.e. laboratory and field research site.
3. Analysis of organizational factors identified at universities and academic institutions.
4. Analysis of organizational factors in external organizations.

#### 6.7.1.3 Section 3 Conclusions

Conclusions are drawn logically from the analysis. The conclusions shall be consistent with what has been analyzed and shall identify the causes of the accident, such as operator error, which had a direct influence on the occurrence of the accident. Next, factors contributing to the accident that may have influenced the operator's action are noted. Latent conditions and events, which were latent prior to the accident and not recognized as deficiencies by the organization that related to the accident, are made visible as accident-contributing factors.

The conclusions use the same confidence terms as in the analysis. In other words, accident-contributing factors described as "probable" in the analysis are also described as "probable" in the conclusions. The source of the conclusions should be clearly identified, and the words should be taken verbatim from the analysis. If a conclusion on the cause has been reached, it should be briefly stated as the cause of the accident. If the cause cannot be determined, the fact should be stated, e.g. "As a result of the accident investigation, it was not possible to determine the cause of the accident."

#### 6.7.1.4 Section 4 Recommended Safety Actions

Safety recommendations are aimed at preventing future accidents, and according to the ICAO (2001), Annex 13 to the Convention on International Civil Aviation defines safety recommendations as follows:

*A proposal of the accident investigation authority of the State conducting the investigation, based on information derived from an investigation, made with the intention of preventing accidents or incidents.*

Safety recommendations are never intended to create a presumption of responsibility on the part of the specific organization or individual who caused the accident and incident. In addition to safety recommendations based on accident and incident investigations, safety recommendations may be issued from a variety of sources, including reports analyzing large amounts of historical data or safety studies on specific issues.

Safety recommendations describe evidence-based safety issues or deficiencies and provide justification for safety actions. These include a specific recipient to ensure that the organization or other entity involved in the accident complies with the safety recommendation. In general, the recipient is the laboratory, department, the university and academic institutions, and other organizations that are related to the accident. If the recipient has implemented the recommended safety measures by the time the accident investigation report is due, the implemented safety measures are excluded from the recommended safety measures. They are then listed under a separate heading, e.g. implemented safety measures.

Accident investigations often identify safety deficiencies not related to the accident. The issuance of safety recommendations for such deficiencies need not be limited to the cause of the accident but must be related to the subject matter of the accident investigation.

Safety recommendations should not increase risk as a result of their implementation and should not increase risk to other equipment or locations in the laboratory not associated with the accident. The effort and cost of safety measures should be balanced against the severity of the risks involved. Before issuing a safety recommendation, it is recommended that the accident investigation team consult with the laboratory involved in the accident on the content of the safety recommendation and revise it to something that the laboratory can implement. This is because, in some cases, the size and financial situation of the laboratory may not allow the recipient to implement the safety recommendations developed by the accident investigation team.

Where safety deficiencies are considered to pose a serious risk to life, property, or the environment, an early warning or interim safety recommendation should be issued rather than waiting for the publication of the accident investigation report to issue a safety recommendation.

If no safety recommendations are made as a result of the accident investigation, the section on safety recommendations may be omitted and replaced by an appropriate heading on safety measures taken (IMO 2014b).

As stated above, the accident investigation report must contain factual information, analysis, conclusions, and safety recommendations. There is a close relationship between these, with the factual information leading to the analysis, the analysis leading to the conclusions, and finally the safety recommendations. The safety recommendation is followed by a follow-up action.

## 6.8 Follow-up Actions After Issuing Reports

Safety recommendations are not mandatory and not binding on the recipients, so some organizations do not take corrective action. In other industries, there are companies that have re-caused fatal or injury accidents by not implementing safety recommendations issued by accident investigation bodies (Japan Transport Safety Board [JTSB] 2011; 2013). Some accident investigation bodies publish annual reports listing the results of the implementation of safety recommendations made with the aim of preventing the recurrence of accidents.

In the case of universities, etc., regular monitoring and review of the implementation of safety recommendations in the laboratory or other location where the safety recommendations have been made by the health and safety committee or EHS to ensure that they are being implemented is an important process to prevent recurrence of accidents.

## 6.9 Conclusion

The accident investigation process includes the elements of data collection, analysis, and report writing. Inadequate data collection can affect the quality of the accident investigation report. This chapter describes the accident investigation process for serious accidents, such as fires, explosions, fatalities, and serious injuries, and for near misses that may have caused these serious accidents.

The on-site investigation of an accident investigation takes place in the immediate aftermath of an accident and involves a wide range of different types and levels of risk. Accident investigation personnel must assess the risks they will encounter before starting the on-site investigation. They can ensure the health and safety of themselves and the accident investigation team by carrying out an ongoing risk assessment, taking into account the following three points: all foreseeable hazards are identified, unacceptable risks are completely eliminated, and the risk assessment process reduces risks to acceptable or ALARP levels.

Hazards include environmental, physical, biological, and material hazards. In particular, if blood and body fluids are left on-site, appropriate PPE for biological hazards must be worn and used.

This book presents an evidence-based approach to comprehensive accident investigation. The importance of evidence is that it identifies the chronological sequence in which accidents occur, supports robust analysis, and justifies recommended safety actions. Evidence falls into four categories: human evidence, physical evidence, documentary evidence, and electronic evidence. The principle of evidence collection is to collect evidence in the order in which it is most likely to be lost. Evidence that is easily lost includes human evidence, damage to the scene immediately after the accident, broken metal surfaces such as equipment and high-pressure gas cylinders, leaked liquids and chemicals, oxygen and certain gas concentrations in the laboratory, traces on the ground, and weather and sea conditions at the time of the accident.

Human evidence is the most volatile form of evidence, and interviews are a means of obtaining human evidence. PEACE, used by national accident investigation bodies and other organizations, is a cognitive interviewing technique that allows a great deal of factual information to be obtained from the interviewee in a short time. The evidence collected needs to be evaluated and sorted. There are five criteria for evaluation: (1) quantity, (2) frequency and regularity, (3) temporal proximity, (4) reliability, and (5) validity.

The RMQMP model, which is specific to accidents occurring at universities, etc. and the ECFCs used by the IMO have been introduced as analysis tools. These analysis tools can be used to identify many contributing factors. Once the investigation and analysis are complete, the report is written. An accident investigation report consists of factual information, analysis, conclusions, and safety recommendations. There is a close relationship between these, with the factual information leading to the analysis, the analysis leading to the conclusions, and finally the safety recommendations.

Follow-up actions complement the safety recommendations. Regular monitoring and review of the implementation of safety recommendations by the health and safety committee or the EHS in the laboratory is an important process to prevent recurrence of accidents.

Universities, etc. are researching cutting-edge, nonpublic technology for innovation. When preparing accident investigation reports, it is necessary to focus only on safety-related processes without disclosing such cutting-edge technology, and to describe details anonymously rather than using personal names to prevent leakage of personal information. Accident investigation personnel must explain that the purpose of safety investigations is not to punish people, but to prevent accidents from happening again, and distinguish them from criminal investigations. In addition, accident investigation reports prepared by universities, etc. must clearly state this.

In order for universities, etc. to collect reliable accident data and systematically prevent accidents, they will need to standardize their methods of accident investigation and analysis. This effort will include standardizing the accident models used as theory, the approaches to accident investigation and analysis, and the content that should be included in accident investigation reports.

## References


Department of Energy (1999). *Conducting Accident Investigations: DOD Workbook*. Washington, DC: DOE.

Fisher, R.P. and Geiselman, R.E. (1992). *Memory-enhancing Techniques for Investigative Interviewing: The Cognitive Interview*. Springfield: Charles C Thomas Publisher.

Fukuoka, K. (2016). Model of marine accidents developed by the Swiss cheese model and recommendations for systematic accident prevention. Doctoral thesis. Kobe University. (Japanese).

Fukuoka, K. (2019). *Safer Seas: Systematic Accident Prevention*. Boca Raton: CRC Press Taylor & Francis Group.

Fukuoka, K. and Furusho, M. (2016a). Relationship between latent conditions and the characteristics of holes in marine accidents based on the Swiss cheese model. *WMU Journal of Maritime Affairs* 15 (2): 267–292.

Fukuoka, K. and Furusho, M. (2016b). Influence of human factors on the occurrence of accidents. *Journal of Maritime Researches* 6: 1–8.

Fukuoka, K. and Furusho, M. (2017). Convergence of the epidemiological and systemic accident models. *Journal of Maritime Researches* 7: 1–11.

Fukuoka, K. and Furusho, M. (2022). A new approach for explosion accident prevention in chemical research laboratories at universities. *Scientific Reports* 12 (1): 3185.

International Civil Aviation Organization (2001). Manual of Aircraft Accident and Incident Investigation: Part 1 Organization and Planning.

International Civil Aviation Organization (2016). Annex 13 to the Convention on International Civil Aviation, APPENDIX 1. FORMAT OF THE FINAL REPORT.

International Maritime Organization (2014a). *Safety Investigation into Marine Casualties and Marine Incidents*, Model Course 3.11, 2014 ed.

International Maritime Organization (2014b). Guidelines to assist investigators in the implementation of the Casualty Investigation Code (Resolution MSC.255(84)), Resolution A 1075(28).

Japan Transport Safety Board (2011). Marine accident investigation report: fatality of a crewmember onboard chemical tanker *KYOKUHOU-MARU No 2*. MA2011-7. (Japanese).

Japan Transport Safety Board (2013). Marine accident investigation report: fatality of a crewmember onboard chemical tanker *KYOKUHOU-MARU No 2*. MA2012-4.

Milne, R. and Bull, R. (1999). *Investigative Interviewing: Psychology and Practice*. Chichester: John Wiley & Sons, Ltd.

Strauch, B. (2004). *Investigating Human Error: Incidents, Accidents, and Complex Systems*. Surrey: Ashgate Publishing Limited.

# 7

# Contributing Factors Found in Accidents

## 7.1 Introduction

Chapter 1 discusses how comprehensive and scientific accident investigation and analysis of accidents occurring at universities, etc. is limited due to the many obstacles unique to universities compared to the aircraft and marine accident investigation sectors. Few accident investigation reports from universities have been published. This chapter analyses published major accident investigation reports, including court decisions, using the risk management and quality management process approach (RMQMP) model. The model is a modification of the software, hardware, environment, central liveware and peripheral liveware (SHEL) model and the Swiss cheese model, as described in Chapter 5, to be applicable to universities, etc. accidents and is presented in Table 5.1. As the epidemiological accident model was found to be applicable to the investigation and analysis of accidents occurring at universities (Fukuoka and Furusho 2022), the RMQMP model is used in the analysis.

The structure of this chapter is as follows: real examples where a number of accident-contributing factors can be identified from the full version accident investigation reports are taken up first, and the accident-contributing factors are explained in terms of each SHEL element. The next group of events listed are mainly incidents collected in cooperation with the Working Group of the Safety and Health Committee of the Faculty of Agriculture, Kyushu University, and the contributing factors classified by each SHEL element, excluding accidents that occurred during laser experiments.

Corrective actions to prevent recurrence of accidents should be based on the relevant higher-level accident-contributing factors (subject to safety recommendations) listed in Table 5.1 in Chapter 5. The analysis of real examples in this chapter is based on eight cases where a number of accident-contributing factors could be identified. The real examples listed include chemical accidents in laboratories, explosives accidents, high-pressure gas accidents, laser accidents, laboratory equipment accidents, diving accidents, river accidents, and mountain accidents.

As noted in Cautions on conducting accident investigations at universities, etc. in Chapter 6, the real examples introduced in this chapter are not expressed in a way that identifies the names of laboratories and individuals.

*Accident Prevention and Investigation: A Systematic Guide for Professionals, Educators, Researchers, and Students*, First Edition. Koji Fukuoka.

## 7.2 Methods of Utilizing Accident Factors

Methods of utilizing the lessons learned from the accident investigation analysis of each case in this chapter are listed in the Lessons Learned section in Chapter 2. Faculty, principal investigator, and environmental, health and safety staff at each university should investigate and check whether the factors listed in the SHEL elements are present in the universities, etc. to which they belong. If the same accident factors are identified, action should be taken to correct the accident factors, and the higher-level factors listed in Table 5.1 can be used as a reference. As mentioned in the introduction to this chapter, the cases were not subjected to comprehensive accident investigation and analysis. However, it can be expected that corrective action on the accident-contributing factors listed in Table 5.1 from the theory of the Swiss cheese model will reduce the likelihood of accidents occurring.

## 7.3 Software/Procedures

### 7.3.1 Accidents During Cleaning of Laboratory Equipment

The following real case (Tokyo District Court 1971) is an example where lack of procedures contributed to the accident. To prevent the recurrence of accidents, procedures must be documented, including the development of a general standard operating procedure (SOP) describing the storage and handling of hazardous chemicals and a specific SOP describing the handling of specific substances such as potassium permanganate. Measures must also be taken, such as the provision of education and training programs for relevant persons and the ongoing monitoring and review of the implementation status of the programs on a university-wide basis.

**Summary of the accident**: On 28 December 1971, a student and a technician came to a university to clean and wash the pipettes, flasks, beakers, and other laboratory equipment used in research experiments for bachelor's thesis in the laboratory. The laboratory was not equipped with a cleaning solution at that time, so they tried to make one with potassium dichromate and concentrated sulfuric acid, but there was no potassium dichromate available. Therefore, the student took the advice of the technician, placed a beaker of solid potassium permanganate on the floor, and gradually mixed in concentrated sulfuric acid by running it down the side of the beaker. During the mixing of the concentrated sulfuric acid in the beaker, an explosion suddenly occurred, resulting in a mixture of flying chemicals, which were scattered all over the student's body, particularly in both eyes and the face, causing blindness and injuries. The explosion was caused by a mixing of concentrated sulfuric acid poured directly onto a large quantity of solid potassium permanganate.

**Procedures at the local workplace**: The student was advised by the technician that potassium permanganate could be substituted for potassium dichromate to make the cleaning solution. The student washed a 2 L beaker in the laboratory with cleanser, wiped its inner walls with filter paper, dried it, and put about 400 g of potassium permanganate into the beaker. When the student started mixing the potassium permanganate and concentrated sulfuric acid, he asked the technician about the mixing ratio of the chemicals and whether they could be mixed directly. The technician instructed him to "try to put a little bit of it in." Therefore, the beaker was placed on the floor, and concentrated sulfuric acid was mixed in little by little by running it down the side wall of the beaker.

**Use of PPE**: No rules had been formulated for the use of personal protective equipment (PPE) and the student was not wearing PPE. At the time of the accident, dust masks, face masks, and goggles were stored in a glass cabinet in the corridor of another laboratory in close proximity to the laboratory.

**Emergency response plan**: At the time of the accident, the laboratory did not have a safety shower to wash off the harmful chemicals in case of exposure, an eyewash facility in case of chemical eye contact, or an emergency response plan.

**Psychological factor**: The technician knew the dangers of mixing potassium permanganate and concentrated sulfuric acid because he had made the cleaning solution himself as part of his bachelor's thesis and assumed that the student knew this too. The technician therefore misunderstood the purpose of the question the student had asked before mixing. The purpose of the question was whether or not solid potassium permanganate and concentrated sulfuric acid could be mixed directly rather than in liquid form.

**Knowledge, skills, and education and training**: The student had no previous experience of making their own cleaning solutions but knew that they were made by mixing potassium dichromate and concentrated sulfuric acid.

**Development and operation of general SOPs and experiment-specific SOPs**: (1) Potassium permanganate and concentrated sulfuric acid are hazardous substances as defined by the Fire Service Act. When these hazardous substances are to be used by students, they shall be stored, handled, and disposed of under the control of the laboratory and the university, but this had not been implemented. (2) The laboratory had not taken measures such as establishing written procedures for the use of hazardous chemicals.

### 7.3.2 Accidents During River Crossings

This accident occurred during a regular educational club activity at a high school where a student was killed while crossing a river (Kyoto District Court 1984). The contributing factors were inadequate approval procedures for changes to the fieldwork plan in the organization, lack of an emergency response plan and general and specific SOPs, and lack of communication led by the teacher, which are similar to contributing factors to other fieldwork accidents. Therefore, measures to prevent recurrence should also be taken to correct these accident-contributing factors.

**Summary of the accident**: In the summer of 1983, student A (first year), a member of the high school mountaineering club, together with leader B (then third year), subleader C (then second year), and student D (then 1st year) left the mountain lodge at around 7 a.m. on 27 July and went up the Shirato River in the Tadami region to their destination, Mt. Maruyama. It started to rain around noon, and at about 5.30 p.m. Student A entered the river to cross the Shirato River but fell over and was swept downstream. At one point, Student A grabbed a tree extending downstream from the riverbank but was swept downstream again. The search was hampered by the river's sudden rise in water level, and Student A was found nine days after the accident occurred.

**Procedures at the local workplace**: On 27 July, at about 7 a.m., Leader B judged that the river had risen a little, but not enough to be dangerous, and that it would be possible to go upstream, so they all set off from the mountain lodge. With Leader B in the lead and Subleader C in the rear, and Student A and Student D in between, they started their trek up the Shirato River at around 8 a.m. The water level was higher than expected, and they continued up the river, sometimes in waist-deep water. In the afternoon, Student A and Student D often fell because they got stuck on rocks at the bottom of the river and their walking speed tended to lag behind. In order to keep up with Student A and Student D's pace, they were made to stand at the front of the group and continued up the river. At around 5 p.m., Student A slipped and ended up in waist-deep water at the back of the group, so they changed their approach and decided to cross to the other side of the river. Leader B, subleader C and Student D went to the other side of the river in that order. Student A started to cross but fell and was swept downstream. Leader B, who was on the opposite bank at the time, asked the other members of the team for help in rescuing him, but they were too far away to contact him. Leader B then returned to the opposite bank and attempted to rescue Student A, who had been swept away. He managed to catch Student A and tried to lift him to shore, but Student A was too heavy to lift and was swept away again, after which he lost sight of Student A.

**Emergency response plan**: (1) No contingency plan had been developed and the action plan had been changed to a three-night itinerary, one night shorter than originally planned. The itinerary was inappropriate in the current weather conditions. (2) When Student A crossed the river, the group split in two, and Leader B attempted to rescue Student A alone.

**Weather, river conditions**: A rainy season front was stagnating in the Tadami region where the accident occurred. At the time of the accident, it was raining and the river was swollen.

**Accumulation of fatigue**: On 27 July, due to impatience caused by the delay in action, the group was fatigued from the long walk up the river in the rain without lunch and adequate rest.

**Psychological pressure**: Psychological pressure built up as a result of the following actions and conditions. (1) The departure was delayed by two days from the original 22 July due to the stagnant rainy season front, but further delay was difficult because the next plan was scheduled. (2) The group left Kyoto for its destination, the Tadami region, on the night of July 24, even though the rainy season front was stagnant, because it was expected that the rainy season would end within two or three days at the latest, as it usually did.

**Communication**: (1) In the High School Mountaineering Club, the advice of the alumnus took precedence over that of the Mountaineering Club advisory teacher, and then the advisory teacher was not informed of a change of plan to a three-night itinerary and departure date.

**Teamwork**: In the Mountaineering Club, the responsibility and authority of the advisory teacher was compromised by the intervention of an experienced alumnus. Teamwork was therefore not functioning.

**Formulation of general SOPs and school safety management**: (1) SOPs for climbing and mountaineering were not formulated and were left to the discretion of site leaders. (2) In developing the action plan, the opinions of alumnus who were not under the control of the school were taken into account. (3) There were no experienced supervisors on-site and the students were attempting to cross a dangerous river on their own in bad weather. These facts show that there was no organized safety management system (SMS) in place.

**Approval of changes to the experimental plan**: A four-night itinerary was initially drawn up by the leader and approved by the school committee as the mountaineering club's summer mountaineering plan. The plan was submitted to the advisory teacher. The plan was later revised to a three-night itinerary, following advice from the alumnus that a three-night itinerary would be sufficient. The change was not reported to the advisory teacher, and the plan was not formally approved as a change of plan.

### 7.3.3 Similar Incidents Related to Procedures

Among the reported accidents and near misses at universities, contributing factors included unclear or undocumented procedures on-site, and documented procedures that existed but were not followed. There have been many cases where general SOPs and training and education have been required as corrective action, such as the following:

- During an experiment, a Pasteur pipette was shaken by hand to extinguish a fire at the tip of the pipette. The burning alcohol then went into the waste bin and caused a fire.

- A student with long hair accidentally burned the hair on their head with a gas burner during an experiment.
- Without checking the label on the chemical bottle, the hydrogen peroxide solution was mixed with hydrochloric acid, thinking it was nitric acid. This produced hydrogen gas, which triggered the fire alarm.
- 1 L of dichloromethane was placed in a triangular flask and transported on a trolley, the triangular flask fell off the trolley, and dichloromethane leaked into the aisle.
- When using a micro high-speed refrigerated centrifuge, the rotor rack was not fitted correctly, and the rotor rack broke, causing debris to fly out of the centrifuge.

## 7.4 Hardware/Equipment

### 7.4.1 Accidents When Handling High-pressure Gas

The following accident is real example where faulty equipment contributed to the accident (Nakao 1991). To prevent a recurrence of accidents, risk assessment of the design and equipment/facilities should be carried out by an expert prior to the start of the experiment. In addition, it is necessary to ensure safety by establishing specific SOPs that specify whether risk reduction is carried out based on the hierarchy of control and whether there are safety issues when the equipment is used.

**Summary of the accident**: On 2 October 1991, during an experiment using plasma chemical vapor deposition (CVD) equipment in the laboratory, a student operated a valve to close it, causing an explosion in the monosilane container supplying gas to the equipment, killing two students and injuring five others in the experiment by blast and flying debris. The explosion ignited town gas and chlorinated organic solvents, causing a fire. The O-ring in the nonreturn valve deteriorated and had gas flowed back, causing the monosilane gas and nitrous oxide to mix and explode. The exploded monosilane and nitrous oxide tanks were connected. The purge line contained a nonreturn valve that was defective at the time of the accident, allowing nitrous oxide to enter the monosilane tank through the purge line and mix, forming a flammable mixture. Pressure fluctuations due to valve operation or static electricity may have ignited the mixture of monosilane and nitrous oxide.

Following this accident, the High-Pressure Gas Safety Act was amended so that consumers of special high pressure gases, including monosilane gas, are now required to notify the prefectural government.

**Equipment conditions related to the accident**: The O-ring of the nonreturn valve was corroded by nitrous oxide, which caused the gas to flow backward. Two design problems were identified. (1) The purge lines for gases at risk of reacting with each other were not separated as follows. The plasma CVD system was supplied with monosilane gas and nitrous oxide gas. The supply lines were independent, but the purge lines were shared, so the monosilane and nitrous oxide tanks that exploded were connected. (2) The mechanism to prevent the backflow of gas with a nonreturn valve was defective. It was found to be impossible to completely prevent backflow of gas with only one check valve, and this nonreturn valve did not function properly for nitrous oxide gas.

**Knowledge, skills, and education and training**: Students were unaware of the hazards of monosilane gas and nitrous oxide and were electronics students unfamiliar with handling these substances.

**Experiment-specific risk control**: (1) The students lacked the knowledge and skills to implement inherently safe design of the equipment, and the laboratory had deficiencies in its management systems to ensure the safety of the equipment. (2) The laboratory should have consulted internal and external experts at the design stage of the equipment and carried out a risk assessment of the equipment to ensure safe design.

### 7.4.2 Similar Events Involving Equipment and Instrument Handling

The cases listed are accidents and incidents that were reported and where the failure to carry out risk assessments on equipment, machinery and tools was a contributing factor to the accident and incident. To prevent accidents, specific SOPs should be developed and implemented for each experiment, specifying that risk assessment should be carried out not only for defects such as cracks in equipment but also for equipment characteristics such as chemical resistance. Risk assessment of equipment should also be specified in the general SOPs applicable to the whole university.

- A glass beaker was broken while being washed, resulting in a laceration of the hand. The beaker was thought to have been cracked before use.
- When a plastic beaker containing a sodium hydroxide solution was picked up, the container cracked, and the solution splashed around. The plastic container had poor chemical resistance to the alkali.
- An off-spec cylinder was installed in the autoclave, causing damage to a piece of equipment during operation.
- The tracking phenomenon of table taps and plugs caused damage and a small fire.

## 7.5 Environment/Atmosphere

During fieldwork, environmental factors play an important role in the safety of researchers and students. The following accident is real example where the environment was a significant contributory factor (Committee for University-wide Accident Investigation into Diving Accidents at the University C 2006). In order to prevent the recurrence of accidents, a risk assessment of the specific area, such as the experimental sea area, should be carried out, and the experimental plan should be formulated after a detailed assessment of the level of competence of the participants in the experiment to cope with the hazards that may occur in the environment. Marine and mountainous areas have their own safety rules or practices. These should be incorporated into specific SOPs, and systematic education and training programs should be established, as individual safety depends on skill. In addition, emergency response plans should always be prepared as mitigation measures, and education and training should be provided.

### 7.5.1 Accident While Collecting Samples at Sea

**Summary of the accident**: On 4 July 2005, a total of six people – a faculty, a researcher, two postgraduate students, a dive guide from a dive shop, and his client – started scuba diving off the coast of Island A at 9:58 a.m. The research team dived for the purpose of sample collection, while the guide and client were separated for recreational diving after landing on the seabed. At approximately 10:10 a.m., the researcher began to surface alone to clear his mask at a depth of 10 m. At approximately 10:20 a.m., the guide joined the research team and noticed that one person was missing, and they all surfaced together when the guide pointed out the missing person. They returned to the water entry point at 10:33 a.m. and searched for the researcher from the shore but were unable to find him. At 10:54 a.m., a fishing boat sailing nearby was informed of the accident by a member of the research team and contacted the police. At 12:23 p.m., the sea rescue team began a search. At 12:46 p.m., the researcher was found lying on his back on the seabed at a depth of 11 m, far from the sampling site. His mask was on his face but half full of water, and his regulator was disconnected from his mouth. The remaining pressure in the tank was 150 bar, but there was no air in diver's buoyancy control unit. Cause of death was drowning.

**Procedures at the local workplace**: (1) The laboratory had not prepared a manual for diving operations. (2) Health checks were not carried out before diving operations, and the suitability of masks and the weight of equipment were not sufficiently checked and verified before and immediately after entering the water.

**Emergency response plan**: (1) An emergency response plan was not prepared; tanks were not tapped with knives and weights were not unloaded for emergency surfacing. (2) There was no emergency communication network, and the laboratory did not contact the emergency services, such as coast guard, but a fishing boat in the vicinity did contact the police. (3) The site was out of mobile phone range, and there was no means of emergency communication.

**Equipment conditions related to the accident**: (1) The scuba buoyancy control device (BCD) broke after the start of the dive due to aging of the rubber nozzle part of the $CO_2$ cartridge attachment. The BCD had been in use for 16 years since its purchase, and the rubber parts had reached the end of their useful life. (2) The mask was purchased shortly before the visit to Island A and was being used for the first time, so it may have been difficult to fit the researcher's face. (3) The tank that was attached weighed twice as much as a normal tank, and the researchers had mistakenly attached 6 kg weights when they had intended to attach 4 kg weights. As a result, the researcher was unable to control the ascent and descent due to the heavy tank and weight.

**Weather, sea conditions and topography**: (1) The area was a dive site with a low underwater transparency of 6–7 m from the sea surface to the seabed and a strong current at the time of the accident. The current flowed from left (south) to right (north). (2) The seabed topography was complex with rocky areas and V-shaped troughs, making it an advanced diving site. (3) The seabed topography made it difficult for the research team to find the researcher, who had sunk to the seabed after the accident, on their way back from the work area.

**Physiological factors**: The researcher had been very busy in the 10 days before the accident. He may have been fatigued from the voyage on the university training vessel, two days of diving work, driving a car to deliver and transport research equipment to the passenger vessel, and traveling by the passenger vessel from Tokyo to Island A.

**Knowledge, skills, and education and training**: (1) A national license was required for diving for work and research, but the researcher did not have a license. (2) Prior to the accident, the researcher had not performed diving work for four years after attending a diving course in a foreign country and lacked diving skill.

**Communication**: (1) The faculty had instructed the research team to buddy up just before entering the water, but this was not fully communicated to them. (2) The research team did not possess underwater slates and had no means of communicating underwater.

**Teamwork**: (1) The faculty was in a position to supervise the research team and ensure the safety of its members. However, the faculty was involved in the sample collection work himself, and it was difficult for him to keep track of the movements of the other three members of the team. The faculty was therefore unable to fulfil his role as supervisor.

(2) The dive guide was originally supposed to check the residual pressure of the tank and monitor the diving behavior of each individual as a safety person. However, on the day of the accident, the dive guide regularly employed by the faculty was replaced due to illness. The faculty did not confirm with the replacement dive guide that he was to perform his duties as safety personnel. As a result, the diving experiment started in the absence of safety personnel. (3) In diving, the buddy system was a method of ensuring safety. As the laboratory practice allowed solo action, the researcher who surfaced to clear the mask was allowed to act alone.

**Laboratory-specific risk control**: Traditionally, the laboratory had adopted a system where the whole research team loosely monitored each other while members carried out their own sample collection activities in order to increase collection efficiency and had not felt the need to form a buddy system. Even during diving operations, independent activities, such as carrying samples, were tolerated, and no diving specific risk management was implemented.

### 7.5.2 Accidents While Traversing the Mountainside

The following accident is a real example of a student involved in mountain accidents caused by surface avalanches (Tokyo District Court 1977). Leaders need to plan their winter mountain trips with sufficient time, food, and fuel, bearing in mind the following ironclad rules and safety issues that need to be taken into account to prevent accidents.

Fieldwork is often conducted in places isolated from society, making it difficult to contact rescue services in the event of an emergency, and in the event of an accident, it is directly linked to danger to the lives of the participants. In order to prevent these situations, the following are some of the issues that organizations and parties should be aware of when undertaking mountain trekking, mountaineering, or winter climbing (Tokyo District Court 1977).

**General Safety Considerations for Climbing and Mountaineering**

- Be accompanied by someone experienced in mountaineering.
- When formulating the mountaineering plan, select the destination, taking into account the gender, skills, and physical fitness of the participants.
- Carry out a site survey beforehand.
- Prepare food and other equipment for worst-case scenarios.
- Carry out medical examinations and have students participate according to their doctor's instructions.
- Consult the mid-term and long-term forecast of the meteorological agencies and postpone or cancel the climb if warnings or advisories are issued.
- If bad weather, such as a storm, occurs at the site, either cancel or change the plan to avoid exhaustion and wait for the weather to improve.
- The instructor or leader should monitor the health of the participants and should not force the itinerary if any of them are tired.

These are ironclad rules for bad weather.

- Generally speaking, when the weather is bad in the mountains, the ironclad rule is to stay put and wait for the weather to improve. In other words, if you act in bad weather, you risk getting lost due to poor visibility and fog, freezing due to physical exhaustion in a windy and rainy environment, and accidents such as slips and falls due to distraction caused by fatigue.

- In snow-covered mountains, the general rule is to walk along the ridge and never cross the mountainside. Traversing a mountainside carries the risk of encountering a surface avalanche, and there is no way to escape the danger of being caught in the surface avalanche.

**Summary of the accident**: The mountaineering club of the technical college, consisting of 10 members in a group including 7 students and a supervising teacher and alumni with extensive mountaineering experience, held a training camp for spring mountaineering in a mountainous area in central Japan, commonly known as the Central Alps. On 30 March 1977, while staying at the mountain lodge A, a snowstorm occurred, and the supervising teacher and alumni decided to cancel their original plan to go from the mountain lodge A to Mt. Komagatake and descend through the mountainside. While descending in bad weather, the group was hit by a surface avalanche, and seven students were crushed to death. According to the testimony of seven mountaineering experts after the accident, the proper way to descend the mountain in such conditions is to stay in the mountain lodge A and descend the ridge trail without risk of avalanche after the wind has died down.

**Procedures at the local workplace**: At about 06:40 a.m. on 30 March, as the snowstorm was still blowing when they woke up, the supervising teacher and alumni decided to abandon the plan and descend through the mountainside, as they felt it was impossible to proceed from the mountain lodge A to Mt. Komagatake as originally planned in such bad weather conditions. At about 00:20 p.m., they left the mountain lodge A, where they were staying. There was a strong wind outside and a considerable amount of fresh snow on the ground. The group followed the forest line along the slope about 100 m below the ridge, almost parallel to the ridge. The route followed the edge of the forest about 100 m below the ridge line on the mountainside, roughly parallel to the ridge line. Near the edge of the forest, the crusted snow surface was piled up to about knee deep, and the group had to move ahead while ploughing snow.

**Emergency response plan**: The mountaineering plan included only one reserve day for the five action days, with no mention of sufficient reserve days or reserve food.

**Weather condition and terrain**: (1) On the morning of 30 March, the wind speed in the area of the mountain lodge A was 15–20 m $s^{-1}$. The wind and snow made it impossible for people to stand upright, and the snow surface was crusted and dangerous. (2) The terrain at the site of the accident had a slope of about 30°, and at a point about 100 m below the ridge line, the wind and snow had been blowing fresh snow since before dawn on 30 March, and the terrain and weather conditions were such that an avalanche was likely to occur. (3) Unlike the winter mountains, the weather in Japan's spring mountains changes in short cycles of a few days, with periods of bad weather followed by periods of good weather.

**Psychological pressure**: The supervising and alumni decided to descend the mountain in bad weather conditions because of the concerns of the collage and the students' parents about a prolonged stay at the lodge and the psychological anxiety it would cause the students.

**Carrying out a risk assessment, including hazard identification**: The supervisor and the alumni forced the students to descend the mountain without a proper risk assessment, even though there was a significant risk of the students becoming exhausted and dying from exposure to the freezing temperature as they crossed the ridge line just below 2000 m in a snowstorm.

**Knowledge, skills, and education and training**: There were two first-year students who had no experience of mountaineering in the snowy mountains; the second- and third-year students also had only one or two mountaineering experiences in the snowy mountains and no skills in snow walking, avalanche safety, and trail work, subjects that needed to be trained during this training camp.

**Communication**: The supervising teacher did not give avalanche warnings to the students, so some of them saw cracks while walking down the mountainside in group, which were not reported to the supervising teacher.

**Response to the accident and near misses**: (1) One year and two years before this accident, there had been avalanche accidents on the slopes of the Central Alps, both of which resulted in fatalities. (2) In the Japanese magazine *Mountain and Valley* published in February 1977, which was on sale at the time of this accident, a special article on the Central Alps in spring pointed out that the slopes, including those near the accident site, were avalanche paths that required caution. (3) It was reported in the newspapers that there were a number of cases of people being rescued by rescue teams because they had stayed in safe places, such as the same mountain lodge where the group was staying, and were waiting for the weather to improve.

**Development and operation of general SOPs and experiment-specific SOPs**: The decision to descend by traversing the slope route greatly increased the likelihood of encountering an avalanche. There is no way to avoid a surface avalanche once people are caught in one. There was no general SOP in place that clearly stated the ironclad rules of mountaineering, mountaineering plans, emergency response plans, and education and training to ensure safe mountaineering, so the decision to stay in the mountain lodge or descend was left to the supervising teacher and alumni on-site.

### 7.5.3 Similar Events Related to the Environment

- A student was noticed looking pale after outdoor training session and when asked, said he had a headache. On suspicion of heat stroke, he was given oral rehydration fluid and allowed to rest.

## 7.6 Operators/Researchers

### 7.6.1 Accidents in Chemical Experiments

The following accident is a real example where a scale-up compound was disposed of in a combustible waste bin, believing it to be inactivated, but reacted with air and moisture, causing a fire (Fukuoka et al. 2020). A similar accident occurred at University A. A summary of the accident and an analysis of the contributing factors can be found in Chapter 1 for reference. To prevent recurrence of accidents, it is necessary to develop and implement general SOPs for scale-up and specific SOPs for experiments, an approval process for changes to experimental plans, and a systematic education and training program.

**Summary of accident**: A researcher was attempting to synthesize the inorganic material CeSiI, but the synthesized sample appeared black and did not match the expected color of CeSiI, so the compound was removed from the argon-filled glove box and thrown into a waste bin containing paper and used gloves, not realizing that the inertization was incomplete. Sparks were emitted from the compound and scattered throughout the laboratory, causing the paper and gloves in the waste bin to catch fire. The researcher soaked the paper and gloves in water to extinguish the fire but realized that a reaction was taking place and extinguished the fire with a powder fire extinguisher, believing that the cerium remaining in the compound was reacting with the water.

**Procedures at the local workplace**: No written procedures had been established for the disposal of the synthesized Ce compound.

**Emergency response plan**: The researcher was able to use the fire extinguisher immediately and appropriately, so there were no human casualties.

**Environmental factors leading to fire, explosions, etc.**: The CeSiI compound decomposes in air and water but does not ignite. However, inert cerium in powder form can easily ignite on contact with air. This suggests that the inert cerium is considered to ignite on contact with air.

**Reduced level of hazard awareness due to familiarity, etc.**: Small amounts of synthetic Ce compounds could be handled without problems and without chemical reactions. The researcher thought that the disposal process would be the same even for 8 g of synthetic Ce compound, which is more than eight times the normal amount, and could therefore be disposed of without problems.

**Carrying out a risk assessment, including hazard identification**: No scale-up risk assessment was carried out. The researcher disposed of more than eight times the normal amount at one time, which may have accelerated the ignition process.

### 7.6.2 Similar Events Related to the Operators

This section addresses the occurrence of accidents and incidents during laser experiments, as documented in the literature (Hashishin 2012). A number of accidents have been reported in which laser light has been directed into the eyes of the experimenter. To prevent the recurrence of these accidents, it is necessary to require the wearing of appropriate PPE, to formulate general and specific SOPs for the local workplace, and to provide systematic education and training. Furthermore, inherently safe design and protective devices, Steps 1 and 2 of the hierarchy of control, are necessary to prevent laser light from entering the human eye during experiments.

- During an experiment in the development study of a 0.5W neodymium-doped yttrium aluminum garnet laser (Nd: YAG laser), the laser light reflected off the glass surface of a wristwatch worn by a postgraduate student and entered his eyes via the reflection. Fortunately, there was no eye damage.
- A postgraduate student found himself with several holes in his clothing while using a carbon dioxide laser ($CO_2$ laser). His eyes were not affected as he was wearing protective glasses.
- While a postgraduate student was aligning the crystal axes during an experiment with Ti: sapphire lasers (titanium-sapphire lasers), the reflected light from the crystal entered his eye. The student suffered an injury that burned the macula of his right eye.

## 7.7 Organizations/Universities

### 7.7.1 Accidents During Magnetic Spectroscopic Experiments

The following accident (Sendai High Court 1986) is a real example of a failure to reconcile university-specific academic freedom and safety assurance and is also relevant to the general SOPs and experiment-specific SOPs.

**Summary of accident**: On 27 March 1975, Assistant A of Institute A at University D was preparing a magnetic spectroscopic experiment on the Zeeman effect under an extremely high magnetic field using an explosive explosion with an electric detonator, under the supervision of a faculty who was also the director of Institute A. Assistant A was working with Assistant B of University D when the explosives filled in the cylindrical detonator suddenly exploded, killing Assistant A instantly by a facial contusion fracture.

The cause of the accident was considered to be an explosion caused by a streamer discharge resulting from an accidental contact between an electric detonator leg wire and a detonator cable. The research team consisted of four members: the faculty, assistant A, assistant B, and assistant C. The research team was replaced by Assistant A, as the assistant professor and an assistant who were originally scheduled to participate were unable to do so due to business trips abroad and poor health.

**Procedures at the local workplace**: (1) The research team did not have a written procedure for magneto-spectroscopic experiments, and safety equipment and local workplace procedures in accordance with the Enforcement Regulations of the Explosives Control Law were not observed. (2) At the time of the accident, it was known that there had been cases of electric detonators exploding without being wired, but the research team continued with their work on the assumption that they would not explode without being wired.

**Equipment conditions related to the accident**: The equipment was required to have fail-safe features, but these fail-safes were not provided in the following ways. (1) An interlock system or alternative device was not installed. The power switch of the electric detonator should have been turned off when a person entered the explosion chamber, and it should have been ensured that no electrical charge was stored in the power supply when the power switch was turned off. (2) No red light, buzzer, or other alarm device was installed to operate automatically while the safety circuit breaker was in the detonation side.

**Environmental factors leading to explosions**: It was raining on the day of the accident and the laboratory was located near the coast, so conductivity was increased by the sea breeze, and if instantaneous high voltage was applied to the electric detonator, a considerable current could flow.

**Accumulation of fatigue**: Assistant A was observed to be fatigued, distracted, and error-prone, as follows: (1) The experiments were carried out from 24 March, and on 26 March, the day before the accident, preparations for the experiments were made until about 23:00 p.m. (2) After carrying out the experiments on the morning of 27 March, the relevant personnel forgot to discharge the capacitor of the detonation power supply in order to set the voltage to zero and to switch the safety short-circuit switch to the short-circuit side. (3) When the experiment resumed in the afternoon of 27 March, assistant A started to adjust the test set-up without checking the safety of these detonation power supply circuits. The ironclad rule was that the wiring should be done after the last and complete adjustment.

**Reduced level of hazard awareness due to familiarity, etc.**: The experiment should have been conducted in accordance with the Explosives Control Law, but the faculty, based on the principle of academic freedom, held the mistaken belief that it was at the discretion of the researcher to ensure safety and failed to ensure safety in accordance with the law. The judicial decision in this accident ruled against the faculty's claims that the academic freedom at the university had been unjustifiably extended to the field of safety measures. The faculty's claims were as follows: (1) Faculty members were also subject to academic freedom in the choice of safety measures during experiments, which was left to the independent decision of the researchers. (2) The safety measures chosen by the faculty were within the reasonable discretion of the researchers based on their primary judgment, and this judgment needed to be respected.

**Carrying out a risk assessment, including hazard identification**: The research team had a high level of expertise in magnetic spectroscopy of the Zeeman effect under extremely high magnetic fields, but they were not experts in handling explosives and did not carry out an appropriate risk assessment.

**Teamwork**: (1) The faculty was the safety supervisor for the entire experiment, but he ignored laws and regulations regarding safety assurance and professional technical standards. (2) The roles and responsibilities of each person were not clearly indicated, and in particular, the person in charge of safety operations after detonation was not designated.

**Formulation and implementation of general SOPs and experiment-specific SOPs**: Procedures based on the Explosives Control Act should have been developed, but the Laboratory Safety Manager was not appointed, and the procedures were left to the faculty on-site, and experiment-specific SOPs were not developed and implemented.

### 7.7.2 Accidents During River Swimming

The following accident (Fukuoka District Court 2018; Accident Investigation Committee at University E 2016) is a real example of the lack of general SOPs, experiment-specific SOPs, and near miss response.

**Summary of the accident**: On 6 September 2016, at 15:00 p.m., a faculty and nine students participating in Program A at university E started swimming across the river from the middle of the river to the right bank (approximately 80 m) near the mouth of River C on Island C. The participating faculty and seven students swam to the right bank side and noticed that two students had drowned from 20 to 30 m from the right bank. One was rescued by the participating faculty using a canoe moored nearby, but Student A was not found in the water. At 15:24 p.m., a neighbor called the police, and at 16:20 p.m., a boat searching 10 m from the right bank and 3 m under water found Student A. He was taken to the hospital at 16:35 p.m. but was confirmed dead at 17:20 p.m. The cause of death was drowning.

**Procedures at local workplace**: The experience in River C itself was not explicitly mentioned in the syllabus or briefing materials for Program A, and students were first encouraged to swim at the site on the day of the accident.

**Use of PPE**: All participants were not wearing life jackets.

**Emergency response plan**: (1) Emergency response equipment, such as a lifeboat, was not prepared. (2) When the faculty and students on the right bank realized that two students had drowned, they initiated a rescue operation and a call for help to the neighborhood but did not immediately contact rescue services.

**Weather, river conditions, and topography**: (1) After the accident, information was obtained that the site of the accident had a complicated current, low water temperature, and was a place where it was forbidden to swim, and that there had been a drowning accident in River C 20–30 years earlier. (2) The depth of the water varied greatly between the right and left banks, with parts of the right bank being very deep, over 3 m at low tide. (3) The stones on the riverbed were slippery and covered with many oyster shells, which could easily fall over when stepped on.

**Physiological factors**: The physical condition of the students was not properly managed as they were not given enough time to rest before swimming. Most of the students departed from the location of University E by night bus the night before the experience and arrived at the port where the ferry arrived and departed early in the morning. The ferry then took about four hours to reach Island C, and the route bus took 40 minutes to reach the accommodation. The experience on River C began 90 minutes later.

**Psychological factors**: The students were under the impression that they could not drown because there had been no warning of the dangers, such as the near drowning of students the previous year, and they were in a psychological state that underestimated the danger.

**Knowledge, skills, and education and training**: As some students swam to the right bank every year, it was expected that some students would start swimming in the year of the accident, but no survey or risk assessment was conducted into the students' experience and ability to swim in the river.

**Response to the accident and near misses**: No analysis of near misses was carried out, although the previous year's report included an account of the student's experience of near drowning.

**Communication**: (1) Information about swimming in River C was provided just prior to the swim, resulting in a lack of communication between the faculty in charge of Program A and students. (2) The introductory course on fieldwork, which consisted of several programs, was taught by 16 faculty members at University E, but there was no sharing of information on safety and topics among the faculty members. (3) The faculty in charge of Program A has extensive teaching experience on Island C and has conducted forestry-related classes and practical training, and no other faculty members raised any points regarding the content of Program A.

**Teamwork**: There was one faculty for every 23 students. In other programs in the introductory fieldwork courses, the number of students per faculty member ranged from a few to seven, so Program A was not structured as a team that could ensure safety.

**Experiment-specific risk management**: When explaining Program A at University E, safety management and safety measures were not adequately explained, and the students were not interviewed or questioned about their swimming skills, previous hazardous swimming experiences and health status.

**Workplace inspections/audits and corrective action**: There was a need to monitor and audit the departments that were carrying out fieldwork, but this was not being done because the SMS was not in place.

At the time of writing this book, there were few accident investigation reports from universities that had investigated and analyzed organizational factors, and in particular, there did not appear to be any investigations into near misses at this time. It is hoped that in the future, comprehensive and scientific accident investigation and analysis will become widespread, and that these organizational factors will also become visible, thereby improving the overall safety of education and research activities at universities, etc. worldwide.

## 7.8 Conclusion

In this chapter, a total of nine real examples of published accidents, including court decisions, are analyzed in the RMQMP model. The factors behind the accidents are then listed in detail using the real examples in the order of SHEL model. It should be noted that a thorough accident investigation will reveal all five elements of the SHEL model as accident-contributing factors.

An accident while washing laboratory equipment in a laboratory was discussed as relevant to software. The student attempted to make a cleaning solution by mixing legally required chemicals, as there was no cleaning solution available in the laboratory to wash the laboratory equipment at the time of the accident. An accident occurred when a technician incorrectly taught a student how to make a cleaning solution based on his own assumptions. In the case of accidents related to workplace procedures, in order to prevent accidents, it is often necessary to document procedures, such as the formulation of specific SOPs indicating the handling of specific substances, to provide training and education programs for those involved, and to make the formulated procedures and training programs available to third parties for verification of their implementation status.

Hardware introduced an accident while handling high-pressure gas. The cause of the explosion was that the monosilane tank that exploded was connected to the nitrous oxide tank. Although a check valve was installed in the purge line, it was not functioning at the time of the accident, indicating a design flaw in the equipment. Under these unsafe conditions, the nitrous oxide entered the monosilane tank through the purge line and mixed with the monosilane to form a flammable mixture. In order to prevent accidents, a specific SOP should be formulated to specify that a risk assessment of the design, apparatus, and equipment should be carried out by experts prior to the start of the experiment to verify that the equipment, etc. has been risk mitigated based on the hierarchy of control and that there are no safety problems when the equipment is put into operation. Risk assessment includes the development and implementation of specific SOPs for the experiment, which specify that the risk assessment should include not only defects such as cracks in the equipment but also characteristics of the equipment such as chemical resistance.

Regarding the environment, two real examples of fieldwork were discussed. In fieldwork, environmental factors play an important role in the safety of researchers and students. In order to prevent accidents, a risk assessment of the specific area, such as the experimental sea area or region, should be carried out, and the experimental plan should be drawn up after a detailed assessment of whether the skills of the participants in the experiment are at a level that allows them to cope with the hazards that may arise in the environment. Offshore and mountainous areas have their own safety rules or practices. These should be incorporated into specific SOPs, and as individual safety depends on skills, a systematic education and training program should be put in place to improve skills in the local workplace. In addition, contingency plans should always be prepared as a mitigation measure and education and training should be provided.

Central liveware includes the psychological conditions of the faculty, researchers, and students conducting the experiment and is an important component of the human factor. Scale-up accidents have been reported in chemical experiments. In real-life examples, a researcher threw a scale-up compound into a combustible waste bin, believing it had been deactivated, and it reacted with air and moisture, causing a fire. To prevent the recurrence of accidents, it is necessary to formulate and implement general SOPs for scale-up and specific SOPs for experiments, formulate approval procedures for changes to plans, and implement systematic training and education programs. For experiments with hazardous chemicals, risk reduction measures based on the hierarchy of control, assuming that accidents may occur, and mitigation measures in the event of an accident must be prepared.

Peripheral liveware primarily includes organizational factors related to accidents at the operator's university or academic institution. Real examples are discussed where the university's unique academic freedom and safety assurances were not in harmony. In the case of the accident during the magneto-spectroscopic experiment, the university should have appointed a director of safety management with authority and responsibility for the safety of the laboratory. The laboratory should have also developed written procedures in accordance with the explosives control law, but the procedures for the hazardous experiment were left to the faculty on-site.

Due to the lack of reports on comprehensive accident investigation and analysis, it has not been possible in this chapter to show accident patterns for each type of accident, e.g. chemical explosions, fires, fieldwork diving accidents, and so forth. It is hoped that in the future, scientific and exhaustive accident investigation and analysis will be carried out at universities, etc., and that the accumulation of accident data will help to clarify accident patterns useful for accident prevention.

## References

Accident Investigation Committee at Kyushu University (2016). Regarding the fatal accident in the "Introduction to Field Science Research: Yakushima Programme". (Japanese).

Committee for University-wide Accident Investigation into Diving Accidents at the Tokyo University (2006). Report on fatal accidents during diving operations to investigate the causes of the accidents and prevent recurrence of such accident. (Japanese).

Fukuoka, K., Bito, S., and Kinjo, S. (2020). What do we need to prevent accidents and incidents? Using the same methodology and establishing a network. Poster presented online at the 7th Asian Conference on Safety and Education in Laboratory (1 December 2020).

Fukuoka District Court (2018). Judicial decision of Fukuoka District Court, Number (ne) 204 of the year 2018. (Japanese).

Fukuoka, K. and Furusho, M. (2022). A new approach for explosion accident prevention in chemical research laboratories at universities. *Scientific Reports* 12 (1): 3185.

Hashishin, Y. (2012). General accidents and hiyari "Chilly" – Hatto "Alarming" during laser operation. *Journal of Japan Society for Laser Surgery and Medicine* 32 (4): 456–462. (Japanese).

Kyoto District Court (1984). Judicial decision of Kyoto District Court, Number (wa) 569 of the year 1984. (Japanese).

Nakao, M. (1991). *Monosilane Gas Explosion at Osaka University*. Tokyo: Association for the Study of Failure. (Japanese).

Sendai High Court (1986). Judicial decision of Sendai High Court, Number (ne) 204 of the year 1986. (Japanese).

Tokyo District Court (1971). Judicial decision of Tokyo District Court, Number (wa) 5085 of the year 1971. (Japanese).

Tokyo District Court (1977). Judicial decision of Tokyo District Court, Number (wa) 9336 of the year 1977. (Japanese).

# 8

# Systematic Accident Prevention

## 8.1 Introduction

When considering only the prevention of accidents in laboratories and field activities, this can be addressed by establishing a safety management system (SMS) in the laboratory. But when a major disaster or accident occurs that involves the entire university, it is necessary to go beyond the framework of the SMS and implement business continuity management (BCM) that provides an integrated response from the entire organization. To determine which events to include in BCM, there is a method of identifying all the possible risk events that could occur in universities, etc., prioritizing their risk levels, and allocating time, budget, and effort to those with higher risk levels. The accidents that occur in universities, etc., as described in Chapters 1 to 7, are calculated as high-priority risk events. In other words, by coordinating and integrating BCM and SMS, the laboratory's response to major disasters and accidents can be effectively managed, and by using these systems, it is possible to take effective measures in the event of natural disasters and major accidents and to mitigate damage in the event of accidents. Figure 8.1 shows the relationship between BCM and SMS.

Section 8.2 describes the methods used to identify risk events throughout an organization, as well as the systems that universities, etc. should systematically implement to prevent the recurrence of accidents and to proactively prevent them from occurring, including damage mitigation measures in the case of natural disasters. In the framework of these systems, the roles and training of administrative offices, departments, and laboratories are described so that they can be applied to the reader's universities, etc. Section 8.3 describes workplace inspections as a function of the Check process in the SMS, while Section 8.4 outlines accident, incident, and near-miss investigations as one of the Check functions. Section 8.5 describes a method for verifying that a system integrating BCM and SMS actually improves the issues identified in the Check process. Section 8.6 presents methods of using data accumulated through accident investigation analysis to prevent the recurrence of accidents similar to serious accidents that have occurred in the past and to prevent the occurrence of frequent accident types.

The approaches presented in Sections 8.2–8.5 are based on the content actually implemented at Okinawa Institute of Science and Technology Graduate University and National University Corporation Kyushu University, where the author was in charge of BCM and SMS. It should be noted that ISO 22301:2019 provides a comprehensive framework, such as the function of the Plan Do Check Act (PDCA) cycle for BCM as a system, but does not specify the details. The approach in Section 8.6 is a modified version of the approach to accident recurrence prevention and proactive accident prevention in the maritime industry published by the author in *Safer Seas: Systematic*

*Accident Prevention and Investigation: A Systematic Guide for Professionals, Educators, Researchers, and Students*, First Edition. Koji Fukuoka.

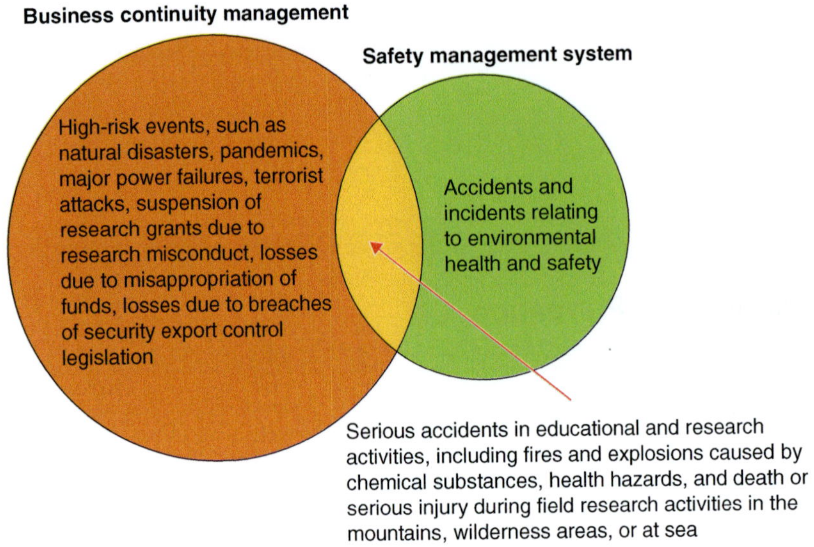


**Figure 8.1** Relationship between BCM and SMS. *Note:* BCM involves the recovery of operations based on a preestablished Business Continuity Plan (BCP) in the event of a natural disaster, pandemic, or other event that has a significant impact on the organization's operations. SMS is solely a method of managing environmental, health, and safety events.

*Accident Prevention*, so that it can be applied to accident prevention in universities and academic institutions.

If an event, such as a fire, explosion, death, or serious injury, occurs during education and research activities, BCM will be used to respond in a systematic manner and resume education and research activities as soon as possible, depending on the level of impact on the organization.

## 8.2 Education

### 8.2.1 Organizational Risk Events

A wide variety of hazards exist in universities, etc., with varying levels of risk. With limited numbers of faculty and staff, annual budgets, and resources, it is the top management who takes the lead in fostering a safe environment and promoting a safety culture, and the proactive commitment of top management has a significant impact on the safety attitudes and behaviors of faculty, staff, and students who are conducting education and research activities on-site.

The following approach is one method of systematically controlling risk events in universities, etc. The Risk Management Committee identifies all risk events and uses a risk matrix and risk reduction principles to reduce and control university-wide risks. By prioritizing risk reduction for events with a high-risk level, this approach can spread the concept and practice of risk reduction throughout the organization, and it is hoped that even faculty, staff, and students who have not previously considered risk or the relative magnitude of risk levels will be able to improve the efficiency of their work by dealing with events based on the priority of the risk level of the event.

Risk events are diverse and include natural disasters, pandemics, major power outages, terrorist attacks, suspension of research grants due to research misconduct, losses due to misappropriation of funds, losses due to violation of security export control legislation, physical hazards caused by chemicals such as explosions and fires, health hazards by chemicals, including health problems due to long-term exposure to carcinogens and toxic substances, accidents such as drowning during field research activities at sea, injuries and deaths due to falls from height, power harassment, etc.

The environmental, health and safety (EHS), or risk management office, leads the process in the Risk Management Committee, and the EHS will be the owner and will oversee and manage risk events related to environmental health and safety. The Risk Management Committee is chaired by the President and consists of the Provost, Vice President, Department Heads, EHS, and others. Administrative experts and faculty members may be appointed as ad hoc members when expertise is required for a particular agenda item. The committee meets regularly, but when a serious event, such as a pandemic or fatal accident occurs, an ad hoc committee is set up to share information, discuss the situation, and decide on policies and countermeasures. It is the university's risk decision-making body.

The risk matrix and the hierarchy of control are used to manage risk events across the university, and the process is as follows. See Chapter 4 for risk analysis, risk matrix, etc.

1. Top management commits at the Risk Management Committee meeting to providing a safe environment and promoting a culture of safety in campuses and disseminates this commitment throughout the organization.
2. The head of the secretariat, department heads, EHS personnel, and others brainstorm about the organization's risks and identify all of them.
3. Decide which department or owner will be responsible for implementing risk reduction measures for the risks identified.
4. The owner takes the identified risks back to the department and analyses them. If statistical data have not been collected, the severity and frequency are determined by qualitative or semi-quantitative methods.

   The selection of the ordinal description of the severity scale can be presented as a scale using the number of casualties as well as the daily operating costs calculated from the university's annual budget amount for one day when the university is closed for one day. For example, consider the risk analysis of accidents that occur during education and research activities. For the severity scale, an explosion and fire occur in a chemistry laboratory, resulting in several injuries, and the laboratory is partly destroyed by the accident and cannot be reopened for several months. For the frequency scale, the records of how often the above accident has occurred in the past at the reader's universities, etc. should be examined. Based on the records, a frequency level is determined, e.g. once every 30 years.
5. Each owner presents to the Risk Management Committee the results of the risk analysis of the events assigned to the owner.
6. The Risk Management Committee discusses the content of the risk analysis, including severity and frequency values for each event, and decides on the risk analysis on which agreement can be reached.
7. The results of the agreed risk analysis are presented in a risk matrix. All risk levels identified in the risk matrix can be compared with other risk events and are visualized.
8. The Risk Management Committee decides how to deal with the risks visualized in the risk matrix and the priorities for dealing with them. When setting priorities, the risk events are arranged in order of the value obtained by multiplying or adding the severity and frequency levels. Creating an Excel spreadsheet with these values makes the ranking of risk levels clear.
9. For events where the result shown in the matrix is A, the risk is in a category that cannot be justified except in exceptional circumstances and the risk is mitigated according to the as low as reasonably practicable (ALARP) principles. It means that the risk is managed according to

the methodology in the hierarchy of control in Chapter 2. The methods include elimination and substitution.

The method of elimination is to stop the research or project itself, i.e. to cancel the initiation or continuation of the research or project covered by the event. Alternatively, substitution is that if there is a lower risk event with the same objectives and expected outcomes, it can be replaced. Therefore, these are the two options. To give an example, suppose it was initially decided to set up a research facility in an insecure location abroad. Research in such a location posed a daily threat to the lives of staff, and the location proved difficult to achieve the research objectives. Risk reduction would be planned either by canceling the establishment of the research facility itself or, if the establishment of the research facility is a priority, by selecting another area where the same objectives are expected to be achieved.

**10.** If the risk levels are B and C, the risk must be reduced based on the concept of the hierarchy of control, but attempts should be made to avoid them as they are unacceptable levels for universities, etc. as described in concept of the risk management process in Chapter 4. If this cannot be avoided, risk reduction can be achieved in the following ways:

Take steps to reduce the level of either or both of the severity and frequency scales. For example, if a chemical explosion is registered as a category C event, the methods to be used by laboratories handling explosive chemicals should be listed in the risk reduction section. If the same experimental results can be obtained, change the material to a nonexplosive chemical. This will reduce both the severity and frequency scales and reduce the overall risk level. If there is no alternative substance as described above, as is often the case, methods to reduce the severity scale should be considered. The severity scale should be reduced to a scale that can be controlled even if an explosion occurs, e.g. by not increasing the amount of chemical used from the mg unit. Particular care should be taken when scaling up, as many explosion accidents have occurred and resulted in casualties when the unit is reduced to g unit. Chemical fume hoods, glove boxes, etc. are used as engineering controls, and blast shields are provided as the risk of explosion has already been identified. Personal protective equipment (PPE) should also be specially designed to be worn when handling explosives. These measures can reduce the severity scale. Another possible method is for a person to remotely control the robot to carry out the experiment. In this case, even if an explosion occurs, people will not be affected, and the risk level will be reduced. Fill in the severity and frequency scales in the columns after the measures have been taken.

**11.** Level D is a widely accepted risk, and the status of whether the risk controls being taken remain effective will be constantly monitored.

**12.** For other risk events, the owner will compile risk reduction measures but will bring them back to the department to plan and implement risk reduction measures. The details of the plan will be reported at the Risk Management Committee meeting.

**13.** When a new risk event occurs, an ad hoc meeting of the Risk Management Committee is held to decide on the risk treatment, ownership, and control of the risk event.

**14.** The above processes allow the entire university's risk events to be arranged in order of priority and visualize risk reduction measures and current risk levels for each event. This provides a bird's eye view of risk events and current risk treatment measures for the entire university and allows for quicker risk treatment. Secondarily, staffing, budget, and allocation of human effort will be optimized by the risk matrix (Table 8.1).

**Table 8.1** Risk management process sheet using the risk matrix (example).

| Rank of risk level | | Severity | frequency | Description of the risk (explain severity scale and frequency scale, if possible) |
|---|---|---|---|---|
| 1 | Earthquake/ tsunamis | 5 | 2 | According to Japanese experts, seismic activity has increased in various parts of Japan since the Great East Japan Earthquake of 2011. In the past 20 years, Japan has experienced three major earthquakes: the 1995 Great Hanshin-Awaji Earthquake, the 2011 Great East Japan Earthquake and the 2016 Kumamoto Earthquake. This Prefecture is no exception. In December 2017, geologists discovered that an area had been hit by four huge tsunamis, with wave heights of 20–30 m, caused by earthquakes. The buildings on the main campus were built to new earthquake-resistant standards, so they are not expected to be damaged even by an earthquake of intensity 6 or higher. In terms of damage from tsunamis, the main campus is located at an elevation of over 30 m and is therefore not affected by tsunamis, but the C campus is located in a low-lying coastal area, and the buildings there are at risk of being completely destroyed by a tsunami. A tsunami of 2–3 m can completely destroy wooden buildings and a tsunami of 5 m can affect reinforced concrete buildings. |

| Risk control (countermeasures) | New severity | New frequency | Progress (monitoring) | Owner |
|---|---|---|---|---|
| If everyone knows how to protect their lives from earthquakes and tsunamis, the damage can be reduced. (1) To protect your own life from earthquakes and tsunamis, get education and training on how to respond to earthquakes and tsunamis. (2) During the day, experience evacuating on foot from your home to the nearest tsunami evacuation site. (3) In the laboratory, secure equipment in preparation for earthquakes and adopt safe methods for storing hazardous materials. (4) Conduct earthquake and tsunami drills for the entire university. (5) Monitor the above procedures. Record the drills in the "Earthquake and Tsunami Drill Records." | 4 | 2 | In year A, the earthquake and tsunami Business Continuity Plan (BCP) was formulated and approved by the University President. After individual training for each task force and laboratory, a comprehensive university-wide drill was conducted in year B. Some issues arose during the comprehensive drill, so these issues were addressed in the BCP and the results were presented to the Risk Management Committee. | EHS |

*Note:* After plotting the severity and frequency scales on the risk matrix, the owner will develop risk reduction measures. The table provides an example of risk reduction measures and regular monitoring. The Risk Management Committee keeps a table of risk events for the whole university. The owner summarizes the risk events for which their department is responsible in an Excel spreadsheet as shown in the table.

### 8.2.2 BCM and SMS

There are two main categories of systems for preventing accidents and recurrences in universities and other institutions and for crisis management in the event of serious accidents and a disaster: SMS and BCM. It is the crisis and safety management experts who create the systems and structures that enable universities and other institutions to formulate and implement SMS and BCM systems. However, to achieve the objectives of preventing the recurrence of accidents and crisis management, it is important that all affiliates, from top management to faculty members, administrative staff, laboratory staff, and students are aware of the objectives of the systems established in the organization and are able to operate them on a daily basis. In other words, to formulate and implement methods for preventing accidents and mitigating damage in the event of an accident at universities, etc., it is essential to provide explanations and training on the system for the administrative departments, faculty members, laboratory staff, and students. Sections 8.2.3–8.2.8 of this chapter explain these topics with specific examples. For more information on SMS, see Chapter 4.

In many universities, etc., all the processes described in this chapter will be seen as new endeavors and will require a great deal of effort to implement. Even when a system has been introduced, the process of operating the system may have its own organizational structure, culture, and climate in universities, etc., and these may become barriers to the operation of the system. If such barriers cannot be removed, creativity and ingenuity will be needed to lower the level of the barriers and to change some of the organizational structure or working practices.

### 8.2.3 Relationship Between BCM and BCP

In the event of a major disaster, pandemic, major accident, or terrorist attack, the operation of universities, etc. will inevitably be interrupted and forced to close temporarily. In such cases, BCM is a system in which the entire university formulates a systematic BCP in advance and effectively and efficiently resumes the operation of universities, etc., based on the plan. BCM includes the following processes:

- The BCP specifies the roles of faculty members and administrative staff by job title and a road map (timetable) for resuming operations of universities, etc. by a specified date. The BCP is developed by experts in BCM and SMS. The role assignments of faculty and staff must be recorded under their individual names in a procedural document that is legally subordinate to the BCP and must be kept up to date.
- The development of the BCP is the Plan process of the PDCA cycle in BCM. The Do process involves the BCP expert explaining the contents of the BCP to top management, the administrative office, departmental administrative offices and laboratories, and conducting drills. Regarding the implementation of drills, for major risk events, such as natural disasters, drills are conducted for each group whenever there is a change of personnel in charge. Comprehensive drills for the entire university, including all administrative offices, departments, and laboratories, should be conducted once a year.
- The Check process involves inspecting the performance of assigned roles in administrative offices, departments, and laboratories. In addition, the level of performance and status of the comprehensive exercises carried out across the university will be reviewed from the perspective of whether they meet the standards of the BCP, and items that fall below the standard will be improved. In other words, if any part of the organization is found to be in a situation where the intended risk mitigation or recovery measures are not proceeding smoothly, even though

the exercise is being carried out in accordance with the plan, the situation will be analyzed and the plan revised accordingly. Furthermore, if a disaster does occur and the BCP has been implemented but the results are not as planned, this Check process includes investigating and analyzing the factors that led to the failure after the event and clarifying the problem areas.

- The Act process involves amending the BCP itself to address issues raised by the Check process, as well as revising the BCP to reflect changes in the environment and societal demands. The revised issues are communicated to the entire university and, if necessary, training is provided for issues related to the revised parts.

In BCM, items that do not correspond to the actual situation are corrected in this way, and when the person in charge of a role is transferred, a new person is appointed to carry out training, etc., and as a result, the method of resuming an organization's operations interrupted by a risk event is continuously improved. The system is designed to enable the resumption of operations of universities, etc., as planned, within a specified period of time, regardless of when or if a major disaster or risk event occurs that triggers the requirements of the BCP.

### 8.2.4 Relationship Between the BCP and Manuals

The BCP establishes task forces in the administrative office for each role, and a road map of what each task force should carry out is formulated. For example, in the BCP earthquake and tsunami, more than 10 task forces are established under the Disaster Response Headquarters, each formed for resuming university operations that have been interrupted by a disaster.

Each task force is assigned disaster-related tasks that are relatively similar to their normal duties. The General Affairs Division is responsible for setting up the Disaster Response Headquarters, which is the decision-making body of the university, and plays a central role in the task forces under the Headquarters. The General Affairs Division organizes and summarizes the disaster operations reported by each task force and reports them to the Disaster Response Headquarters. It also receives information from Departmental Affairs offices, which are in charge of the administrative work of the laboratories belonging to each faculty. Facilities Division is mainly responsible for assessing damage to buildings and land, running emergency generators, and repairing or replacing the laboratories themselves. Human Resources and Academic Affairs Divisions are responsible for collecting information from faculty, staff, and students to confirm their safety and for providing support to those affected. Environment, Health and Safety Division is responsible for the handling of controlled substances, hazardous substances, and laboratory animals. Health Office is responsible for the management of injuries, psychological support, and hygiene for faculty, staff, and students. Information Systems Division is responsible for the recovery and backup of IT equipment. Finance Division is responsible for the management and distribution of stockpiled materials and equipment, including water and food.

The details of the issues for which each of these task forces of the Administrative office is responsible are to be defined by each task force in a separate manual. The following is a sample of the disaster operations in the BCP earthquake or tsunami of Facilities Division. (1) List of personnel and their detailed respective roles in Facilities Division in the BCP; (2) procedures for starting up emergency generators; (3) determination of the order of priority for the supply of power to university buildings, laboratories, etc. by means of emergency power supplies; (4) power supply agreements with local utilities to secure fuel for backup generators; (5) determination of building inspection priorities and safety inspection procedures in the performance of building safety inspections; (6) procedures for repairing or replacing damaged laboratories, classrooms, offices, etc.; (7) assessment of damage to buildings and land and amount of damage.

If there is a change of faculty or staff, the names of those with roles in the BCP must be changed and the list of roles updated. The manuals of each task force may be amended with the approval of the head of that task force and are reviewed several times a year. The BCM is reviewed at meetings, such as the Risk Management Committee, attended by the President and other top management, and the draft text is approved, which requires fewer changes than the manuals. Information security must be strictly enforced in the various task force manuals and departmental office manuals in which individual names are listed.

For each task force manual, the expert who formulated and built the BCM and SMS (hereinafter referred to as the "BCM expert") presents the skeleton of the manual to each task force and departmental office, and each task force and departmental office adds details adapted to their department to create the BCM procedures. In this way, manuals can be formulated efficiently and adapted to each department.

### 8.2.5 Safety-related BCM

This section outlines the safety-related BCPs that fall under the Plan process of BCM. This is only a list of key points that organizations should consider, but the BCP must also clearly identify the task force and personnel responsible for implementing these points, as well as the means and methods of implementation.

#### 8.2.5.1 BCP Earthquake and Tsunamis

The BCP earthquake and tsunamis defines the criteria for the strength of the earthquake that triggers the BCP. The main issues to be addressed by the organization in the event of earthquakes and tsunamis above those criteria occurring in the area where the universities, etc. are located are as follows. In addition, the BCP earthquake and tsunamis should include information on two scenarios: one where infrastructure, such as electricity, water, and gas, is available following an earthquake or tsunami, and one where the infrastructure has been destroyed: (1) take actions to protect one's own life; (2) confirm the safety of family members; (3) rescue the injured on campuses; (4) send safety confirmation information to the university; (5) establish the Disaster Response Headquarters and hold campus-wide meetings; (6) decide on and publicize the temporary closure of the campus; (7) implement emergency response and emergency priority tasks; (8) gather safety confirmation information on faculty members, staff, and students and follow-up on casualties; (9) check the backup status of information systems; (10) check for damage and deal with fires and gas leaks; (11) inform the local population of any leaks of chemical or radioactive substances that may cause health hazards and implement measures to prevent leaks; (12) handle interviews from the media; (13) assess the damage to buildings and land, including offices, lecture halls, laboratories, and research equipment; (14) identify dangerous areas on campus; (15) set up facilities to receive people who cannot return home; (16) provide meals, necessities, etc.; (17) provide material and financial support to victims, including faculty, staff, and students; (18) provide psychological support to victims; (19) dispose of debris, etc. on campus; (20) formulate and carry out repair work on buildings and laboratories after assessing the damage; (21) plan for the resumption of education and research activities; (22) receive and manage relief supplies; (23) repair and replace damaged laboratories and lecture rooms; (24) apply for budgets to restore facilities and equipment; (25) resume education and research activities; and (26) disband the Disaster Response Headquarters.

#### 8.2.5.2 BCP Tropical Cyclone

According to the World Meteorological Organization (2024), tropical cyclones are the second most dangerous natural hazard after earthquakes. Tropical cyclones often bring strong winds, torrential rain, storm surges, flooding, tornadoes, and lightning, posing a significant threat to life and property. The names of these meteorological phenomena vary according to the ocean in which they occur: Typhoon is the name given to those that form in the Northwest Pacific; hurricane to those that form in the Caribbean Sea, Gulf of Mexico, Northwest Pacific, and North Atlantic; cyclone to those that form in the Bay of Bengal and the Arabian Sea; severe tropical cyclone to those that form in the Southwest Pacific and Southeast Indian Ocean; and tropical cyclone to those that form in the Southwest Indian Ocean.

Unlike earthquakes, advances in meteorological technology and the analysis of meteorological data by meteorological organizations worldwide mean that universities can predict the date and intensity of tropical cyclone impacts several days in advance. This means that the BCP Tropical Cyclone can be activated before a tropical cyclone hits the university. For example, the BCP may be activated if the Meteorological Agency predicts that a tropical cyclone with wind speeds of 41 m/sec or greater, or the one of sufficient intensity to cause a power outage lasting three days or more, is heading for the area where the university is located. If the campus is located in a coastal or low-lying area, preparations should be made for flooding from storm surges or breaches in river levees. If there are casualties or damage to buildings caused by a tropical cyclone, the same measures should be taken as for casualties or damage to buildings caused by earthquakes and tsunamis.

The main issues to be addressed by the organization in the event of a tropical cyclone are as follows: (1) test emergency generators and prepare for the power outages by Facilities Division; (2) check storage conditions to prevent leaks of hazardous substances; (3) back up research data, etc. in case of a power outage; (4) relocate critical equipment and data from flood-prone laboratories to a location that is not susceptible to flood damage; (5) establish the Disaster Response Headquarters and decide on temporary campus closures; (6) ensure that contractors and other workers on campus are aware of safety measures; (7) take shelter in a safe place away from windows during a storm warning; (8) survey damage to buildings, land, roads, etc. after storm warnings are lifted and confirm safety on campus; (9) after confirming the safety on campus, resume education and research activities; (10) disband the Disaster Response Headquarters.

#### 8.2.5.3 BCP Fatalities and Injuries in Education and Research Activities

The requirements for triggering the BCP vary according to the judgment of universities, etc., but may be defined as the occurrence of one or more fatalities or one or more serious injuries. The faculty and laboratories should cooperate with the Accident Investigation Committee established by the Emergency Response Headquarters to clarify the actual circumstances of the accident and formulate measures to prevent its recurrence. The main objectives of this BCP are to mitigate damage, prevent secondary disasters, resume experiments suspended due to the accident as soon as possible, and prevent the accident from recurring.

The main issues to be addressed by the organization in the event of a fatal or serious injury accident during education and research activities are as follows: (1) notify the emergency services of the accident by the research group at the scene at the time of the accident; (2) implement first aid and secondary disaster prevention measures; (3) notify the university's crisis management office or relevant department via the emergency contact network; (4) notify the families of victims of the accident; (5) establish the Emergency Response Headquarters and decide on responses, such as suspending experiments related to the accident; (6) set up the Accident Investigation Committee by the Emergency Response Headquarters and initiate an accident investigation; (7) send university staff to the scene and report the situation at the accident site to the Emergency Response Headquarters; (8) deal with media coverage of the accident; (9) submit an accident investigation report by the Accident Investigation Committee to the Emergency Response Headquarters; (10) decide the measures to prevent recurrence by Emergency Response Headquarters in response to the contributing factors of the accident in the report; (11) disband the Emergency Response Headquarters; (12) continue to support the families of the victims; (13) follow up with the recipient by the Health and Safety Committee or EHS until the organizational factors and safety recommendations related to the laboratory are fully implemented; (14) restart experiments that had been suspended after corrective actions had been completed.

#### 8.2.5.4 BCP Major Fire

The requirement for triggering the BCP can be defined as when initial firefighting measures fail and the fire has spread to other laboratories and hallways. If a fire results in death or serious injury, the Accident Investigation Committee should be established with external experts based on the BCP fatalities and injuries in education and research activities. The Accident Investigation Committee must conduct a comprehensive accident investigation and analysis, separate from the fire investigation by the fire and police authorities, to clarify all local workplace and organizational factors and formulate measures to prevent a recurrence of such an accident.

It is recommended that universities, etc. communicate regularly with local fire authorities and conduct regular fire drills, evacuation drills, and portable fire extinguisher use drills using university laboratory buildings. Portable fire extinguishers have limitations. If flames from a fire reach the ceiling of the room, the fire cannot be extinguished because it is beyond the limit of what a portable fire extinguisher can extinguish. In a house in Japan, it takes about two minutes for flames to reach the ceiling. In a laboratory, it may take less than two minutes for flames to reach the ceiling because, unlike ordinary houses, flammable and combustible materials are stored in the laboratory. In Japan, it takes about 10 minutes for the fire brigade to arrive at the scene of a fire, although this time varies from country to country. In addition, firefighters do not start extinguishing the fire immediately. In the event of a fire in a laboratory, the local fire brigade will ask the researchers who know the situation about the type, quantity, and location of hazardous or water reactive substances stored in the laboratory. The reason for this is to avoid explosions or the spread of fire through water-based firefighting. Therefore, it must be recognized that to save the lives of people in the laboratory and to protect important data, samples, and laboratory animals, it is important that the person who detects the fire and their colleagues use portable fire extinguishers to extinguish the fire in its early stages.

In the BCP Major Fire, the main subjects that the organization will address are as follows: (1) notify the fire brigade of the fire; (2) evacuate faculty, staff, and students from the building where the fire occurred; (3) take the names of all the members at the evacuation site; (4) establish the Disaster Response Headquarters and decide to temporarily suspend research in the affected laboratories; (5) cooperate with the fire brigade and provide it with information about hazardous materials and water reactive substances stored in the laboratories where the fire occurred; (6) extinguish the fire by the fire brigade and assess the damage; (7) establish measures to prevent a recurrence based on the cause of the fire as determined by the fire brigade and the results of the accident investigation conducted by EHS; (8) repair the affected laboratories or set up alternative laboratories; (9) resume research activities in the affected laboratories; (10) disband the Disaster Response Headquarters.

### 8.2.6 BCM and the Laboratories

This section describes how the organization's BCM relates to the laboratory. Because of the differences in what is studied between the arts and humanities and science and engineering, the response to emergencies, such as earthquakes, tsunamis, tropical cyclones, storm surges, and floods will be different in part. Below is a list of responses for science and engineering departments that store chemicals and are particularly at risk from fire. In the arts and humanities, responses can be prepared by deleting subjects that do not apply. The emergency priority work list, safety confirmation, and damage assessment are issues common not only to the laboratories but also to the administration and the departmental offices.

#### 8.2.6.1 Creation of an Emergency Priority Work List

People are in a state of panic, unable to think and act rationally when a disaster or other emergency occurs. Laboratories will be in chaos. To avoid such a situation, plan in advance what the laboratory will prioritize in the event of an emergency such as earthquakes, tsunamis, major power failures, or flooding due to tropical cyclones. The laboratory will then develop a list of emergency priorities and communicate it to affiliates and others in the laboratory so that they can access the list and take action in the event of an emergency.

In the emergency priority work list, enter the names of the laboratory personnel, their normal duties, the emergency priority tasks, and the time required to complete each priority task listed. The time required should be entered, taking into account cases when infrastructure is available and cases when infrastructure has been completely destroyed.

Priorities would include the following: securing human life; confirming safety of laboratory personnel; protecting laboratory animals; protecting radioactive materials, toxic gases, and other substances that would have a significant impact on the environment if released; taking initial measures in the event of these materials or organisms escaping from the laboratory; taking measures in the event of a power failure; responding to a fire; and assessing damage to research equipment.

#### 8.2.6.2 Preparing for Power Outages

Power outages may be caused by earthquakes, tropical cyclones that disrupt the electrical grid, or by disruption of the balance of supply and demand for electricity from power suppliers. Researchers need to be prepared for situations where they may be unable to enter their laboratories for long periods of time because access to the building has been restricted as a result of safety inspections by the Facilities Division following earthquakes or tropical cyclones. The impact of a prolonged power failure on research needs to be carefully considered and countermeasures taken. The following is a sample checklist of actions that a laboratory storing chemicals should take in the event of a power failure.

- Is there a risk of critical equipment being damaged by the power failure? If there is a risk, take steps to shut down the equipment.
- How will the power failure affect ventilation in the laboratory? Take measures if ventilation is stopped.
- Is there a risk that the power failure will cause an ongoing chemical reaction to get out of control? If there is a risk, take steps to stop the process.
- Is there a risk that important experimental data could be lost as a result of the power failure? If there is a risk, take possible measures, such as backing up data.

- Will a power failure adversely affect experimental plants and animals? If so, take measures such as using the uninterruptible power supply (UPS) and the laboratory's own power supply or relocating the laboratory.
- Are procedures in place to safely terminate experiments that have been unattended for long periods before evacuation? If so, follow the procedures before evacuation.
- Are fridges or freezers used to store unstable chemicals/compounds connected to a UPS or similar device? Take measures to control runaway reactions of these chemicals during a power failure.
- If access to the building is by ID card, has the Facilities Division given instructions on how to enter the building in the event of a power failure that disables the card function?

#### 8.2.6.3 Confirming the Safety of Faculty, Staff, and Students

Confirming the safety of faculty, staff, and students after a major disaster is the most important issue. Disaster Response Headquarters set up at each university, etc., collect safety confirmation information via the Internet and IT equipment. In the safety confirmation of staff and students belonging to laboratories, requests for follow-up examinations of those who have not responded or those who have responded as seriously injured are sent from the administrative office to the person in charge of the laboratory via the departmental administrative office. The request is sent by email and includes information on the current location, whether and to what extent injuries have occurred, and the need for a rescue call. The laboratory manager must follow up on the contents of the survey. The content of the follow-up is eventually reported by the departmental office to the Disaster Response Headquarters, which follows up on the safety confirmation information. If rescue or other assistance is requested from the sender, it will be handled in cooperation with the emergency services. The laboratory should maintain a contact list of its members and emergency contacts to facilitate follow-up, and this list should be updated whenever there is a change in personnel.

#### 8.2.6.4 Recording Damage to Research Equipment

When research equipment is damaged, the damage and the amount of damage must be reported to the administrative office. In the case of a major earthquake, the government will usually classify the disaster as a major disaster in accordance with the law, and a certain amount of damage will be compensated accordingly. The format of the disaster report to be submitted to the relevant ministry, if any, should be followed, but generally, the damage to each piece of equipment and an estimate of the amount of damage are recorded. To facilitate the reporting of damage, it is recommended that a hard copy contact list of relevant companies, including equipment suppliers or manufacturers, be prepared.

If a laboratory is damaged, the repair and replacement work will be listed in the BCP's disaster recovery process chart and managed by the university as a whole. Depending on the extent of the damage, the laboratory may not be able to resume research activities on campus within the timeframe specified in the BCP. In such cases, it may be necessary to collaborate with another university and carry out research there, so it is recommended that this is taken into account during normal times.

#### 8.2.6.5 Preparing for Damage Mitigation Measures in the Event of a Major Disaster and Accident

Once a major disaster or accident occurs, the damage and impact on universities, etc. can be enormous. However, it is possible to minimize the damage by preparing for the worst in a BCP during normal times and by learning and training how to reduce the damage to people and minimize the damage to laboratories in the event of a major disaster and accidents. The following topics are those for which the laboratory's faculty, staff, and students should be prepared on a daily basis as damage mitigation measures and include, as a minimum, the necessary emergency response training. These fall under the Plan and Do process of BCM.

#### 8.2.6.5.1 *Preparedness for Emergencies*

- In the event of an occurrence, such as an earthquake or fire, there must be sufficient space in the laboratory for people to take action to save their lives. Evacuation routes and exits must be free from obstacles that would prevent people from evacuating to exits. Sufficient space for life-protecting action includes space for a safe route for people to evacuate if the platform supporting the equipment, such as a laser, moves during an earthquake.
- In the event of a major accident, information must be rapidly disseminated via the emergency communication network to the fire service, rescue services, and university personnel. At the accident site, measures need to be taken to mitigate damage and to prevent secondary disasters. These measures include the provision of first aid and other assistance to those affected, as well as the implementation of measures to prevent colleagues and others from being involved in the accident.
- Emergency safety equipment, such as safety showers, eyewash units, carbon dioxide fire extinguishers, powder extinguishers, and spill kits, have to be clearly marked and unob structed on the route to them. Researchers need to be trained to rush to the emergency safety equipment in the event of accidental exposure to chemicals on the face or body, or in the event of a fire, and to use the equipment to take damage mitigation measures.
- Two evacuation routes from the laboratory to the designated evacuation site must be provided, illustrated, and displayed in the laboratory. In the event of a fire, the two routes are to be used in an alternative manner. The route from the laboratory to the emergency exit of the building that can be reached in the shortest distance shall be the first route. It is recommended that laboratory personnel undertake a walk-through of the evacuation route from the laboratory to the evacuation site at a time of year when staff-student turnover occurs, for example, in October.
- The most common cause of injury to people during an earthquake is injury accidents caused by being hit by or being trapped under objects that have fallen or toppled over due to the shaking of the earthquake. It is therefore important to look around the laboratory and, if there are any, such objects, to take measures to prevent them from falling or toppling over.
- Tidiness in the laboratory, including laboratory tables, needs to be practiced. The maintenance of a tidy laboratory environment is beneficial not only in terms of reducing the risk of accidents occurring but also in preventing the loss of materials and documents, contamination, and the spread of fire.
- The responsible person and emergency contact details in case of emergency are displayed on the aisle-side door at the entrance to the laboratory.
- In the event of an emergency, such as a fire, evacuation, or injury requiring first aid or cardiopulmonary resuscitation (CPR), the laboratory should establish an emergency response plan. In buildings where automated external defibrillators (AEDs) are installed, laboratory personnel should be trained in AED use and CPR. Training in CPR and AED use can be obtained from fire or rescue services.
- In the event of a fire, it is imperative that portable fire extinguishers compatible with the chemicals stored in the laboratory are in place and that personnel are trained in their use.

Carbon dioxide fire extinguishers are used in laboratories due to their ability to extinguish fires without causing damage to surrounding research equipment. Carbon dioxide is heavier than air and accumulates on the floor, which means that it does not affect people during normal firefighting in a standing position. However, when used in a confined, small room, care should be taken to ventilate the room as there is a risk of oxygen deficiency.

- In the event of a chemical leak or spill, it is essential to differentiate between instances where large quantities of chemicals are released and those where smaller quantities are involved. This distinction is crucial in determining the appropriate response as it determines whether the incident can be handled by laboratory faculty, staff, and students, or requires the expertise of chemical specialists. Additionally, the response method must be tailored to the specific characteristics of the leaked or spilled chemical. In certain instances, highly toxic chemicals may necessitate the deployment of specialized units capable of handling such chemicals from the fire service.
- When conducting field research activities in rivers, oceans, wilderness areas, or mountainous regions, the risk of being caught up in natural disasters, such as earthquakes, tsunamis, flooding due to storm surges, or landslides due to torrential rain, should be assessed and clearly stated in the experimental plan. First, to avoid being caught up in a natural disaster, a hazard map showing the vulnerability of the experimental area, meteorological and oceanographic data, and records of past natural disasters are studied in detail to identify hazards and assess risks. Experiments are then avoided in areas of high risk and unacceptable risk.

  It is important to note that even if an area is assumed to be low risk based on historical data or other factors, it may still be involved in the natural hazards listed above due to significant changes in the surrounding environment or global climate change. Therefore, it is crucial to prepare for such situations by implementing damage mitigation measures.

  In the event of inundation due to a tsunami, storm surge, or flood, stay in a sturdy reinforced concrete building on a floor above the expected inundation level and wait there until the warning issued by the meteorological agency is lifted. The best places to evacuate in the event of tsunamis are high hills, and local governments usually designate evacuation sites and publish them on hazard maps. Information on expected tsunami and flood levels can be found on the websites of meteorological organizations. If this information is not available, it will be possible to distinguish between dangerous and safe areas in the experimental area by searching and studying news reports of past major disasters in detail.

  If a torrential downpour occurs and the local government or meteorological agency issues a flood warning, evacuate during the daytime when it is still daylight to a nearby high point or a safe place if one exists. At night, stay on the highest floor of the building in which you are staying, as there are many potential hazards on the streets when moving around.

  It is also important to have access to local news at all times. For research activities in unfamiliar areas, it is recommended to be accompanied by a local guide who is familiar with these natural hazards and evacuation centers.

#### 8.2.6.6 Training

- It is recommended that emergency drills be conducted whenever there is a change of staff or students in the laboratory. The emergency response plan covers a range of potential disasters, including earthquakes, fires, evacuations, power failures, chemical spills, and radioactive releases from laboratories. Depending on the laboratory, if a unique event is expected, an additional emergency response plan for that event is added and training is provided.

- After conducting various types of training, the content of the training should be recorded and kept in a training record book. The content of the training record should include the date and time of the training, the content of the training, the participants, the reason for the training, and any other issues or problems identified during the training, as well as any improvement actions. The next training session should incorporate these improvements.

### 8.2.7 BCM and the Role of Administrative Offices and Laboratories

If there is a major, widespread power failure in the area where universities, etc. are located, or if there is a power failure due to a natural disaster, such as an earthquake or tropical cyclones, the university's power grid will fail and laboratories will be without power for several days or longer. In administrative offices, PCs and other IT equipment that depend on electricity, as well as lighting and other facilities, will be unavailable. In the laboratories, there may be damage to critical equipment, loss of experimental data, and damage to laboratory animals and plants. The lack of power to research equipment could cause fires and explosions due to runaway chemical reactions, and people could be injured in secondary disasters resulting from these accidents.

When such an abnormal situation occurs, it is the administrative office that systematically leads the resumption of operations based on the BCP road map. The BCM activities of the administrative office are described in Section 8.2.5 and are therefore omitted. Under the administrative office, the departmental administrative offices and laboratories will carry out recovery activities while dealing with disaster operations and emerging issues within the limited infrastructure. To ensure that these activities can be carried out smoothly, the administrative office and the departmental administrative offices should develop the BCM, emergency priority work list, and power outage countermeasures in their respective departments.

As stated in Section 8.2.6, the laboratory will operate on the basis of the BCM, but it is necessary for the faculty to appoint personnel and formulate the emergency response plan in preparation for events that occur or are anticipated within the laboratory and to conduct training when personnel changes occur due to transfers.

### 8.2.8 Briefing and Training of the Administrative Office and Laboratories

In order to achieve the desired accident prevention and mitigation through a harmonious combination of the BCM, managed by the administrative office, and the emergency response plan, mainly implemented by the laboratories, the office in charge of the BCM must provide explanations and training on BCM to the top management of the organization, as well as to the administrative and laboratory staff. As a prerequisite, the BCM needs to be approved as an official document. Approval as an official document of universities, etc. ensures that the commitment of the President and the application of the BCM to the entire university are known and disseminated to all faculty members, staff, and students and that the downstream processes of the Plan process, such as education and training, audits, and implementation of improvements, can proceed without delay.

1. Once the official documents have been approved, the priority is to brief and train the administrative office. The BCP establishes a number of role-based task forces in the administrative office under the umbrella of the Disaster Response Headquarters, which play an important role in making the BCM work.
2. The departmental briefing will take place after the administrative office briefing has been completed, but will focus on an overview of the BCM, the role of the department, and the

relationship and responsibilities between the department and the laboratory. In the BCM, the departmental administrative offices sit between the laboratories and the administrative office and have the role of acting as a bridge between the two. The departmental administrative offices work with BCM experts to assist the laboratory in developing emergency response plans and methods for disaster response and accident prevention. The BCP briefing to the departmental administrative offices and the training are one and the same, and the office in charge of BCM should provide a sample manual listing the items and procedures that the departmental administrative offices should be trained on, and it is recommended that the training be conducted within about a month of the briefing. As can be seen from Ebbinghaus's forgetting curve, people forget the content of explanations over time, and the effects wear off. The recommendation to conduct the training within approximately one month of the briefing also applies to the various task forces of the administrative office.

3. The laboratory briefing will take place after the departmental administrative office briefings have been completed. The process can be effective and efficient if the explanation and question-and-answer session takes place at the Faculty Assemblies, the decision-making body of the faculty. In a university with more than 20 000 faculty, staff, and students, it would probably take more than two months to explain the process to all Faculty Assemblies. After briefing the Faculty Assembly on BCM as a whole and giving an overview of the training to be provided in the administrative office and departmental administrative offices, Section 8.2.6 should be explained, together with an acknowledgement of the important role of the university for the faculty members who manage the laboratories.
4. It should be mentioned in advance that the BCM expert will carry out an audit of the processes according to the BCM regarding the maintenance of the minimum required training and manuals as outlined by the administrative office, the departmental administrative offices, and the Faculty Assemblies. Details of the audits are explained in Section 8.3.

## 8.3 Workplace Inspections and Internal Audits

The workplace inspection, which is common in many countries in universities, etc., is a statutory inspection under the Occupational Health and Safety Regulations, and its purpose is to ensure that the work of employees and the workplace environment are safe. The SMS workplace inspections described in Chapter 4 of this publication include inspections under the Occupational Health and Safety Regulations and aim to prevent recurrence of accidents and health damage and to take proactive accident prevention measures. The SMS workplace inspections are carried out on a regular basis through the Health and Safety Committees set up in each department.

In universities with more than 20 000 faculty, staff, and students and a large number of laboratories, it is physically difficult to inspect all the laboratories within a year, and some ingenuity is required. One method is to estimate the level of risk according to the characteristics of the substances stored or handled in the laboratory, the presence of substances that affect human health, and the nature of field research activities, and to prioritize the order of workplace inspections according to this level of risk. This approach uses the risk matrix presented in Chapter 4 and prioritizes the highest risk groups of laboratories. For the reasons given above, it is possible for small universities to carry out SMS and BCM workplace inspections at the same time, but for large universities it is recommended that these workplace inspections are carried out separately.

### 8.3.1 BCM Workplace Inspections

BCM workplace inspections are conducted by a BCM expert to review the emergency response plans and disaster preparedness in place in each laboratory. The results of BCM workplace inspections lead to improvements in the sustainable education and research activities of the laboratories and contribute to the improvement of BCM in the university as a whole.

Inspection items for BCM workplace inspections covering the laboratory are listed in Section 8.2.6, but not all are listed. Inspection items for BCM workplace inspections targeting administrative offices and departmental administrative offices are listed in Sections 8.2.4 and 8.2.8. The BCM expert should ensure that each task force manuals, departmental administrative office manuals, training record sheets, and the names of the persons responsible for the roles assigned in the manuals are checked to ensure that they are up-to-date and effective.

### 8.3.2 SMS Workplace Inspections

In general, SMS workplace inspections are carried out by a team consisting of EHS staff and specialists, including faculty and technical staff. Depending on the discipline of the workplace inspection laboratory, this may include faculty members from the same discipline. One approach is to outsource workplace inspections to private contractors. However, taking into account that the EHS has the best knowledge of the actual safety situation in the laboratory and that the EHS conducts accident investigations and analyses and takes measures to prevent recurrence of accidents and incidents, it is more effective for the EHS to take the initiative to carry out such workplace inspections. The EHS is kept informed of the safety situation in the laboratory. The EHS can make appropriate changes to the subjects and content of preaccident workplace inspections and make the laboratory much safer by analyzing why accidents and near misses occurred after the accident.

It is recommended that the EHS clarify the purpose, criteria, and items to be inspected during the inspection and that the chairperson of the Health and Safety Committee participate in the inspection. SMS workplace inspections should be conducted in the presence of the laboratory faculty and the PI. The inspector's side should provide safety information, including accidents and incidents at the university and updates on regulations. The laboratory side provides information on experimental procedures, including specific SOPs, explains the storage conditions for hazardous materials, the status of compliance with legal and university policies and guidelines, including ventilation and exhaust, and actively communicates with the inspectors. In this way, communication between the two sides is kept active. If unsafe conditions are discovered on-site, instruct laboratory personnel to take immediate corrective action if it is possible to do so on-site. Those requiring more time should be discussed with the laboratory faculty to establish a timetable for corrective action.

Inspection items for high-risk chemicals include but are not limited to the following: These contribute to laboratory safety and are a way of ensuring that chemicals do not fall from their shelves, leak, ignite, and cause a fire or explosion in the event of an earthquake in seismic areas.

- All chemicals are stored separately according to their hazard category and incompatibility prohibition. The National Academy of Science (2011) provides the following examples. Organize and store chemicals in categories such as flammables, strong acids, strong bases, peroxides, etc. Organize things so that everyone knows where everything is. This is important in the event of a fire. The first thing firefighters will do when they arrive on the scene is to find out if there are any people trapped in the laboratory or building where the fire has broken out, as well as the type, location, and quantity of any hazardous materials stored in the laboratory.

- Chemicals are organized and illustrated to ensure proper labelling (CAS number, molecule number, quantity, date of acquisition, storage location, name of person responsible, etc.), storage method (dark, cold, etc.), and which chemicals are stored where.
- Laboratories where chemicals are stored must be labeled according to the Globally Harmonized System of Classification and Labelling of Chemicals with a sign on the side of the door aisle.
- Chemicals should always be purchased and used in minimum quantities.
- Laboratory procedures should clearly state that experiments with chemicals that pose a risk of fire or explosion should not be carried out alone. It is difficult to deal with an emergency alone. It is also hoped that by conducting experiments with two or more people, errors and mistakes can be avoided during the experiment.
- Peroxide formations should be dated on receipt and again when the container is opened so that the substance can be disposed of according to recommendations such as those in the safety data sheet (SDS). The content of SDSs is mainly industrial in nature and may vary depending on the publisher of the SDS. In addition, the hazards are only listed in general terms, and there is no information on which substances are particularly hazardous, so it is important to be aware that there are limitations on how they can be used in a laboratory (National Academy of Sciences 2011).
- Chemicals are not stored in the laboratory fume hood, on the floor, in hallways, or at doorways.
- Flammable liquids are stored in approved flammable liquid containers and storage cabinets. The law specifies the designated quantities, which are quantities that can be stored, and the local fire authority's instructions are followed.
- If flammable chemicals are stored in a room, the chemical storage area is a spark-free environment, and only spark-free tools are used in the room. Special earthing and bonding must be installed to prevent static charging when solvents are injected.
- An inventory of chemicals is carried out at least once a year, and a table or diagram is kept in the laboratory showing what is where, including the room number where it is stored and where it is stored within the same room.
- Processes from procurement to disposal of chemicals are controlled, including legal storage and disposal methods. Procedures must be developed and followed for the disposal of chemicals due to the transfer of personnel, including the closure of laboratories.
- Chemicals are transported in leak-proof containers or chemical transport carts, and freight elevators are used. If a freight elevator is not installed, ensure that no other persons are present when the lift is brought in.
- Laboratory-grade explosion-proof refrigerators and freezers are used. Household items are not used as they are a source of ignition.
- Maintenance and inspection of laboratory equipment is recorded. This includes items that require self-inspection as well as those subject to statutory inspection.
- Guidelines for wearing appropriate PPE according to the hazard level of the experiment have been developed in the laboratory and are monitored and reviewed, e.g. peer review by the research group, to ensure that the appropriate PPE is worn before the experiment begins.
- Compressed gas cylinders are secured with two chain bands and capped when not in use.
- Machine tools are equipped with safety devices or emergency stops and guards.
- Homemade equipment has risk reduction measures according to the hierarchy of control.

- Eating and drinking are not allowed in the laboratory, and eating and drinking areas are separated.
- Legal limits for lighting and noise are observed.
- Legal environmental standards are met.
- Safety-related information, such as accidents and near misses, issued by the EHS is available to laboratory faculty and staff, and meetings of laboratory personnel are held regularly so that lessons can be learned in the laboratory.

## 8.4 Investigation of Near Misses and Accidents

This section deals with the SMS from the point of view of preventing the recurrence of accidents that occur in the organization and proactively preventing them from happening in the first place. It is necessary to establish a centralized reporting system for accidents that occur in universities, etc. The designers must therefore create a system that is easy for researchers to use. For example, the system should ensure that this system can be accessed directly from one of the icons displayed on the university's website and that the website provides accident investigation topics in the software, hardware, environment, liveware (SHEL) elements listed in Table 5.1. Depending on the severity of the accident, the accident investigation report will be either a full version or a simplified version, and the level of investigation and analysis of the events will depend on the severity of the accident.

Serious accidents and serious near misses must be created in a full version accident investigation report by means of a scientific and comprehensive on-site investigation of contributing factors at the local workplace and organizational factors at universities, etc., visualizing all latent conditions that have been latent for a long time. On the other hand, incidents that do not have serious consequences should be created in a simplified version. Considering the human resources of the EHS that investigates the accident, the timeframe for the accident investigation, and the other tasks of the EHS, it would be physically difficult to use the approach described in Chapter 6 to carry out an accident investigation analysis and to produce a full version report in all cases. The appropriate approaches are listed below. See the core of the SMS in Chapter 4 for more information.

1. Serious accidents, such as fires, explosions, fatalities, and serious injuries, shall be investigated and analyzed by a third-party committee or the EHS accident investigation personnel. The investigation and analysis are carried out using the approach outlined in Chapter 6, including interviews with those involved in the local workplace and the organization, and are based on a comprehensive evidence-based investigation and analysis of all the SHEL elements listed in Table 5.1. The accident investigation report is completed in full version.
2. Near misses that may have resulted in a serious accident shall be investigated and analyzed by the EHS accident investigation personnel using the same approach as in (1) and a full incident investigation report shall be prepared.
3. Accidents other than serious accidents shall be reported by the accident investigation personnel of the EHS by filling in the SHEL elements in Table 5.1. Interviews with the persons involved are conducted by the EHS. A simplified version of the accident investigation report is prepared, consisting of a summary of the accident, the SHEL elements in Table 5.1 and measures to prevent recurrence.
4. Incidents, other than near misses, which may result in a serious accident, should be reported by the laboratory personnel by completing the SHEL elements in Table 5.1. Interviews with

the persons involved are conducted by the laboratory personnel. A simplified version of the accident investigation report is prepared, consisting of three parts: a summary of the accident, the SHEL elements in Table 5.1, and measures to prevent recurrence.

5. The full and simplified versions of the accident investigation report should be made available for use by the EHS, risk management office, university top management, the Health and Safety Committee, and the laboratory faculty.

## 8.5 Follow-up Action

In BCM, the follow-up action is to check whether the issues identified in the BCM workplace inspection have been corrected and are being operated. The deadline for completion of corrective actions is set by month and date. Follow-up inspections must be carried out periodically to ensure that the actions taken during education and training are implemented in a sustainable manner. Follow-up includes a visit by the BCM expert to the local workplace to verify that the BCM instructions are being implemented in the laboratories. Inspections will focus on "BCM and the laboratories" described in Section 8.2.6. For administrative offices and departmental administrative offices, the BCM procedures and the emergency priority work list are the subject of follow-up.

In the SMS, a follow-up action means an on-site inspection to ensure that the action for the deficiency to be corrected has been carried out and that the documentation associated with the deficiency has been appropriately corrected. The implementation of the corrective action is decided at the end of the accident investigation, in consultation with the laboratory that caused the accident and the relevant departments. If the adoption of inherently safe design measures also involves the ventilation system of the whole building, this has to be discussed with the Facilities Division. Where changes to general or specific SOPs are required in relation to organizational factors, efforts should be made not only to amend the relevant documents but also to ensure that the corrective items are implemented in the laboratory or at the local site of the field research activity. If the organizational factors require changes to the overall organizational structure of the university, this may take a long time as many departments are involved and administrative procedures are followed.

The results of the follow-up will be reported to the Risk Management Committee and the Health and Safety Committee, and the follow-up will continue until the corrective action in the laboratory or organization has been completed.

## 8.6 Using the Data

This section shows how the data collected in the accident or incident investigations and analysis can be used. As a comprehensive set of accident-contributing factors has been identified at the accident investigation and analysis stage, this data can be processed in Excel spreadsheets to present the main accident-contributing factors for each type of accident (Fukuoka 2016; Fukuoka and Furusho 2016).

In particular, if the data obtained from the full version of the accident investigation reports are collected and processed, it is possible to retrieve from the data not only the accident-contributing factors but also the process in the experiment that caused the accident and the methods to prevent it by allowing each laboratory to view the accident factor data for each type of accident. By using this data in education and training programs, faculty, staff, and students will be able to learn more specifically about risk assessment, risk management, and SMS through case studies, and the lessons learned will contribute to improving safety and fostering a safety culture in the organization (Fukuoka and Furusho 2022).

Figure 8.2 illustrates the process of using data to prevent accidents. The figure is a modified version of a set of processes for preventing accidents from recurring in marine industries published by Fukuoka (2019), adapted to accidents that have occurred in universities and other academic institutions.

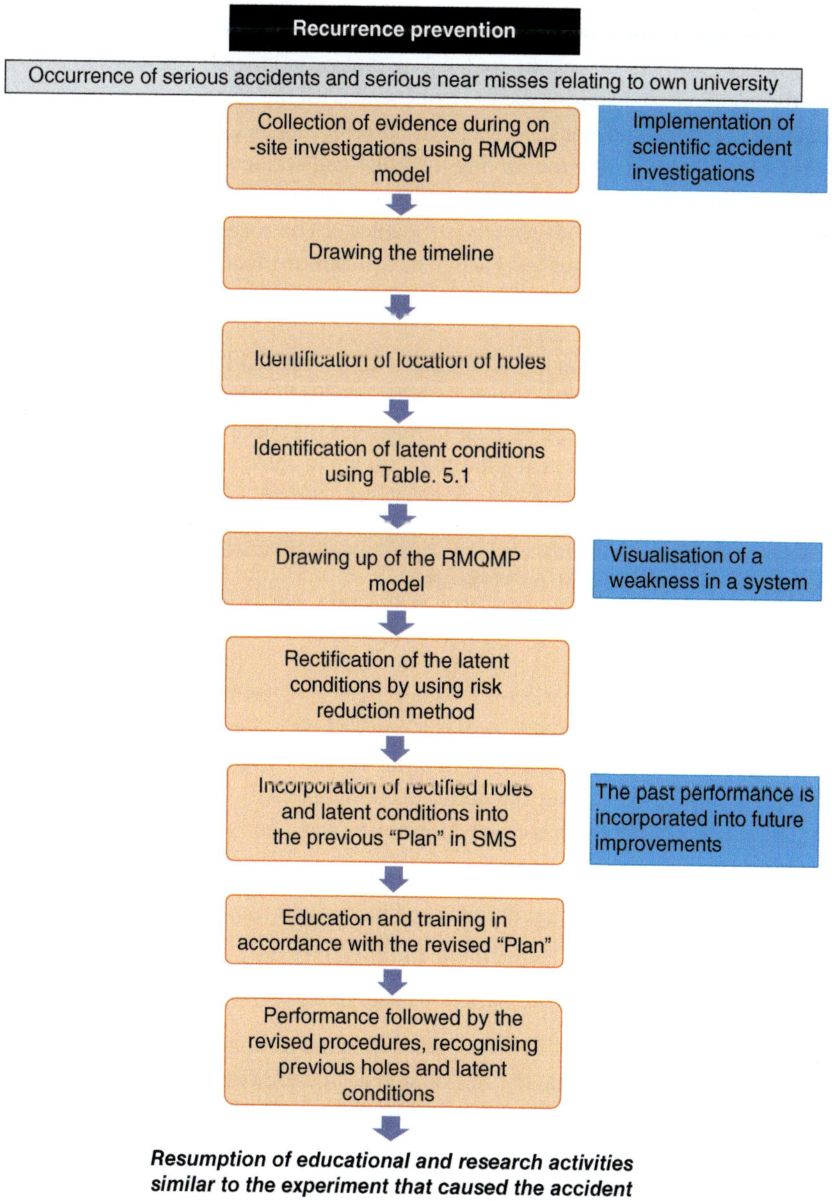


**Figure 8.2** Process for preventing recurrence of accidents using data. *Source:* Adapted from Fukuoka (2019).

### 8.6.1 Prevention of Accident Recurrence

The following steps should be taken to prevent the recurrence of serious accidents such as fire, explosion, death, or serious injury. This procedure can also be applied to nonserious accidents. It is described here on the assumption that a comprehensive and scientific accident investigation is carried out.

1. When investigating accidents in universities, etc., use the risk management and quality management process approach (RMQMP) model to collect all evidence related to the accident.
2. Use a timeline diagram or ECFC to show the sequence of events in chronological order.
3. Use the PDCA cycle to show the status of implementation of the SMS used by the organization and the risk management process in the local workplace to identify any holes in the system.
4. Identify the accident-contributing factors associated with the accident event according to the classification of latent conditions, i.e. accident-contributing factors, given in Table 5.1.
5. Draw the RMQMP model after identifying the holes and accident-contributing factors.
6. To prevent a recurrence, correct all holes and latent conditions that caused the accident using the hierarchy of control methods shown in Chapter 2. The corrective actions should refer to the measures listed in the higher-level accident-contributing factors of Table 5.1.
7. Incorporate the corrected holes and latent conditions in the Plan process of the SMS at the time of the accident and revise the SMS.
8. Conduct education and training in accordance with the revised Plan process of the SMS.
9. When resuming education and research activities similar to the experiment that caused the accident, special attention should be paid to the holes that occurred during the risk management process. In addition, the latent conditions shown in Table 5.1 should be checked again to ensure that they have been rectified in advance before the experiment is resumed.

### 8.6.2 Proactive Accident Prevention Measures

The data collected through the investigation and analysis of all accidents at the universities, including accidents at other universities that have been made public, can be used to proactively prevent accidents. In the Lessons learned section of Chapter 2, the analysis of the accidents at Universities A and B led to the conclusion that although the types of accidents were different, many of the accident-contributing factors were common. In other words, universities, etc. can reduce the risk of accidents occurring by correcting the common accident-contributing factors (Fukuoka and Furusho 2022). The procedure for proactively preventing accidents is as follows:

1. Reanalyze full version accident investigation reports published on accidents at other universities, etc., to determine the location of holes and factors contributing to the accident. This data will be collected separately as "Data from other universities."
2. Identify the types of accidents or incidents that occur frequently from the statistical data of your university or academic institution.
3. For the types of accidents identified in 2, determine the location of the holes and the factors contributing to the accident from the full version accident investigation reports.
4. Compare the hole locations and contributing factors identified in 3 with those identified in the accident investigation reports of the other universities, etc. in 1.
5. If, as a result of the comparison in 4, there are locations of holes and accident-contributing factors not mentioned in the accident investigation report of the university to which you belong for the type of accident that frequently occurs, and these should be added.

6. Draw the RMQMP model incorporating safety recommendations from the own and other universities once the holes and accident-contributing factors from 5 have been identified for the most common accident types.
7. Information on proactive accident prevention measures for the most frequent accident types should be disseminated by EHS to all laboratories. The use of this information in systematic education and training programs for faculty, staff, and students will also increase the effectiveness of accident prevention.

There is a method of reducing the risk of a particular type of accident by quantifying the factors contributing to that type of accident from accumulated statistical data and calculating the main factors contributing to the accident. A Pareto chart is used to quantify accident-contributing factors. This makes it possible to identify contributing factors that occur frequently in certain types of accidents. There are limits to the resources available to an organization, including people and budgets. The whole organizational accident prevention process is based on the idea of "vital few, trivial many." In other words, as the process of accident prevention methods progresses, the focus is initially on the frequent accidents, and the number of such accidents decreases. After that, it is hoped that the number of less frequent accidents will also decrease (Fukuoka 2016; Fukuoka 2019).

Considering that the larger the amount of scientific data, the more accurate the results of data analysis, it is recommended that data be collected in cooperation with other universities, etc., in order to prevent accidents before they occur.

## 8.7 Conclusion

Laboratory responses, including disasters and major accidents, can be effectively managed by coordinating and integrating BCM and SMS and using these systems to prevent accidents before they occur and to take effective mitigation measures when they do occur. One method of systematically controlling all risk events related to universities, etc. is to identify all risk events and use a risk matrix and risk reduction principles to reduce and manage organizational risks. This approach allows risk reduction concepts and actions to be spread throughout the organization by prioritizing risk reduction from the highest risk events. The risk matrix and the hierarchy of control are used to priorities the management of university-wide risk events, and effort and resources are invested in higher-level risk events. The higher-level risk events are managed in a systematic manner as part of BCM.

In the event of a major disaster, pandemic, serious accident, or terrorist attack, the operation of universities, etc. will be interrupted and forced to close temporarily. In such cases, BCM is a system that enables the university, etc. to resume operations effectively and efficiently based on a preestablished plan (BCP) to systematically restore the organization's operations. BCM is based on the PDCA cycle, where any items that are not in line with the actual situation are continuously corrected, and when the person in charge of a role changes, a new person is appointed and trained, so that the method of resuming the organization's operations after a risk event is continuously improved. The system is designed to enable the resumption of operations of universities, etc., within a specified period of time as planned, regardless of when and if a major disaster or risk event occurs, which is a prerequisite for triggering the BCP.

It is the crisis and safety management experts who formulate the BCM systems and SMSs and build the mechanisms to make them operational. However, in order to achieve the objectives of disaster control and accident recurrence prevention, it is important that all affiliates, from top management, and faculty members to administrative staff and laboratory personnel and students, are aware of the objectives of the systems established in the organization and are able to apply them on a daily basis. In other words, in order to formulate and apply methods for preventing accidents and mitigating damage in the event of an accident in universities, etc., it is essential to provide explanations, education, and training on the system to them.

Approaches to maintaining and improving the above systems include workplace inspections, internal/external audits, and accident and near miss investigations. SMS workplace inspections include inspections in accordance with the occupational health and safety regulations and also aim to prevent the recurrence and avoidance of accidents and health hazards. Under the auspices of the Health and Safety Committees set up in each department, regular SMS workplace inspections are carried out. SMS workplace inspections are conducted in the presence of a faculty member or PI in the laboratory. The inspector's side provides safety information, such as accidents and incidents and updates on regulations. The laboratory side provides information on experimental procedures, including specific SOPs and storage of hazardous materials to ensure active communication. If unsafe conditions are identified on-site, instructions are given to take immediate corrective action if it is possible to do so on-site. Those requiring a longer period of time should be discussed with the laboratory faculty to determine a schedule for corrective action.

BCM workplace inspections are conducted by BCM specialists who review the emergency response plans and disaster preparedness in place at each laboratory. The results of the BCM workplace inspections lead to improvements in the sustainable education and research activities of the laboratories and contribute to the improvement of BCM in the university, etc. as a whole.

It is necessary to verify that the items identified during the SMS and BCM workplace inspections have been corrected and are operational in the laboratories. Regular follow-up must be carried out to ensure that the identified unsafe conditions are corrected and that the corrected conditions are implemented in a sustainable manner. Follow-up involves the EHS or a BCM specialist visiting the local workplace to verify that the SMS and BCM findings are being implemented in the laboratories.

The use of collected data and accident prevention are closely related. In particular, when the data from the full version of the accident investigation reports are collected and processed, the data will provide not only the accident-contributing factors but also the process in the experiment used by each laboratory to cause accidents and the methods to prevent them. By using these data in education and training programs, faculty, staff, and students will be able to learn more specifically about risk assessment, risk management, and SMS through case studies and can use the lessons learned to help improve safety and promote a safety culture in their organizations.

These important issues have not been implemented in many universities and academic institutions worldwide. However, the author believes that if these issues can be resolved, there will be a paradigm shift in the safety of laboratory and field research activities.

## References

Fukuoka, K. (2016). Model of marine accidents developed by the Swiss cheese model and recommendations for systematic accident prevention. Doctoral thesis. Kobe University. (Japanese).

Fukuoka, K. (2019). *Safer Seas: Systematic Accident Prevention*. Boca Raton: CRC Press Taylor & Francis Group.

Fukuoka, K. and Furusho, M. (2016). Relationship between latent conditions and the characteristics of holes in marine accidents based on the Swiss cheese model. *WMU Journal of Maritime Affairs* 15 (2): 267–292.

Fukuoka, K. and Furusho, M. (2022). A new approach for explosion accident prevention in chemical research laboratories at universities. *Scientific Reports* 12 (1): 3185.

ISO 22301: 2019(en). *Security and Resilience-Business Continuity Management Systems-Requirements.* Geneva: International Organization for Standardization.

National Academy of Sciences (2011). *Prudent Practices in the Laboratory: Handling and Management of Chemical Hazards.* Washington, DC: The National Academies Press.

World Meteorological Organization (2024). Tropical cyclone. https://wmo.int/topics/tropical-cyclone (accessed 28 July 2024).

# Abbreviations

| | |
|---|---|
| ATC | air traffic control |
| ATSB | Australian Transport Safety Bureau |
| BCM | business continuity management |
| BCP | business continuity plan |
| BRM | bridge resource management |
| CHP | chemical hygiene plan |
| COLREG | Convention on the International Regulations for Preventing Collisions at Sea, 1972 |
| CRM | cockpit resource management |
| CSB | US Chemical Safety Board |
| DHS | United States Department of Homeland Security |
| ECFC | event and contributory factors charts |
| EHS | environmental health and safety |
| FMEA | failure mode and effect analysis |
| FTA | fault tree analysis |
| GEMS | generic error modelling system |
| GHS | Globally Harmonized System of Classification and Labelling of Chemicals |
| HFACS | human factors analysis and classification system |
| ICAO | International Civil Aviation Organization |
| IMO | International Maritime Organization |
| ISM Code | International Safety Management Code |
| ISO | International Organization for Standardization |
| JTSB | Japan Transport Safety Board |
| MAIB | Marine Accident Investigation Branch |
| NTSB | National Transportation Safety Board |
| OSHA | Occupational Safety and Health Administration |
| PI | principal investigator |
| PPE | personal protective equipment |
| QMS | quality management system |
| RMQMP model | risk management and quality management process approach model |
| SDS | safety date sheet |
| SHEL model | software, hardware, environment, liveware model |
| SMS | safety management system |
| SMSM | safety management system manual |
| SOP | standard operating procedure |
| TQM | total quality management |

# Index

Note: Page numbers in *italics* and **bold** refers to figures and tables respectively.




*Accident Prevention and Investigation: A Systematic Guide for Professionals, Educators, Researchers, and Students*, First Edition. Koji Fukuoka.

Printed and bound by CPI Group (UK) Ltd, Croydon, CR0 4YY
17/09/2026
14964314-0001